Astroglossary
Revised Edition

by G. Cyr, C.S.D, T.D., S.Ed.D., B.A., M.A.

Agora Books
Ottawa, Canada

Astroglossary: Revised Edition

c. 2019 by G. Cyr

All Rights Reserved. No part of this book may be reproduced, stored in a retrieval system, or transmitted in any form or by any means, electronic or mechanical, including photocopying, recording, or otherwise without the expressed written consent of The Agora Cosmopolitan.

Care has been taken to trace ownership / source of any academic or other reference materials contained in this text. The publisher will gratefully accept any information that will enable it to rectify any reference or credit in subsequent edition(s), of any incorrect or omitted reference or credit.

Agora Books
P.O. Box 24191
300 Eagleson Road
Kanata, Ontario K2M 2C3

Agora Books is a self-publishing agency for authors that was launched by The Agora Cosmopolitan which is a registered not-for-profit corporation.

ISBN: 978-1-927538-34-0

Photo credits:

Front cover is a "collage" of the following images herein described:

1 - "The Blue Marble" is a famous photograph of the Earth taken on December 7, 1972, by the crew of the Apollo 17 spacecraft en route to the Moon at a distance of about 29,000 kilometres (18,000 mi). It shows Africa, Antarctica, and the Arabian Peninsula.

NASA/Apollo 17 crew; taken by either Harrison Schmitt or Ron Evans - https://web.archive.org/web/20160112123725/http://grin.hq.nasa.gov/ABSTRACTS/GPN-2000-001138.html (image link); see also https://www.nasa.gov/multimedia/imagegallery/image_feature_329.html

Public Domain:

File: The Earth seen from Apollo 17.jpg
Created: 7 December 1972

2 - The following is a description of original and supplementary images by ESA/Hubble which comprise this "collage":

(i) One of the largest Hubble Space Telescope images ever made of a complete galaxy is being unveiled today at the American Astronomical Society meeting in San Diego, Calif.

The Hubble telescope captured a display of starlight, glowing gas, and silhouetted dark clouds of interstellar dust in this 4-foot-by-8-foot image of the barred spiral galaxy NGC 1300. NGC 1300 is considered to be prototypical of barred spiral galaxies. Barred spirals differ from normal spiral galaxies in that the arms of the galaxy do not spiral all the way into the center, but are connected to the two ends of a straight bar of stars containing the nucleus at its center.

Credit:

NASA, ESA, and The Hubble Heritage Team (STScI/AURA)

(ii) The star-forming region NGC 3603 - seen here in the latest Hubble Space Telescope image - contains one of the most impressive massive young star clusters in the Milky Way. Bathed in gas and dust the cluster formed in a huge rush of star formation thought to have occurred around a million years ago. The hot blue stars at the core are responsible for carving out a huge cavity in the gas seen to the right of the star cluster in NGC 3603's centre.

Credit:

NASA, ESA and the Hubble Heritage (STScI/AURA)-ESA/Hubble Collaboration

(iii) An illustration [image in yellow] shows the newly discovered planet, Fomalhaut b, orbiting its sun, Fomalhaut.

This structure is a Saturn-like ring that astronomers say may encircle the planet. Fomalhaut also is surrounded by a ring of material. The edge of this vast disk is shown in the background as the curving cloud-like feature that appears to intersect the 200-million year-old star. Fomalhaut b lies three billion kilometres inside the disk's inner edge. The planet completes an orbit around Fomalhaut every 872 years.

Credit:

ESA, NASA and L. Calçada (ESO)

Back cover:

G. Cyr and Constance Cyr

Printed in Canada

This book was published due to the continued support of my wife Constance, and the technical advice of my friends Damian Costea and Andre Tremblay, to whom I will be forever in debt for their generous help with their time and expertise.

Prologue

This book is meant to help you if you feel like dabbling in this wonderful domain of astrophysics. I am trying through this publication to "ease up" your entrance into this new, wonderful world which is literally infinite. I want to share with you the result of many, many hours and nights that I spent in libraries. Instead of searching all over for the terms you're looking, you may find this word or expression right here in this book.

As a conclusion with this "Astroglossary" of mine, I am trying to provide a link between regular people (like me) who are interested in astronomy, and the "ivory tower". Of course, although this book may look like an overly ambitious endeavour, I believe that it as a step in the right direction. I consider it as my modest contribution to the general effort for knowledge and understanding.

The author

Index

A 8
B 15
C 21
D 31
E 38
F 44
G 48
H 54
I 60
J 64
K 65
L 67
M 71
N 77
O 81
P 84
Q 93
R 95
S 100
T 114
U 120
V 122
W 125
Z 128

Aarseth–Brewington– A comet discovered in 1989 with its peak visual magnitude of +2.4 as observed from December 1989 to January 1990.

AAS – American Astronomical Society.

AAVSO – American Association of Variable Star Observers.

Abell catalogue – a catalogue of clusters of galaxies.

Aberration — a distortion of an image due to imperfections within the optical system.

Ablation– smoothing out the surface of a celestial object.

Absolute magnitude – a measurement of a celestial object's luminosity in logarithmic units.

Absolute space – a space devised by Newton; a portion of the sky without any interference by matter or any form of energy.

Absolute zero – the lowest temperature that is theoretically possible in which particles remain motionless. It's value is zero on the Kelvin scale, equivalent to -273.15°C.

Absorption lines – spectral lines in a continuous spectrum resulting from the absorption of light by the atoms and/or molecules composing the absorbing medium.

Accelerated expansion – increase in the speed of the stretching of space.

Acceleration – rate of increase in speed.

Acceleration of the Universe – the increase in the rate of expansion of the universe.

Accretion – the accumulation of material onto a massive object by its gravitational attraction.

Accretion disk – a gravitationally-bound disk of material in orbit around a massive object that feeds the central object via accretion.

Achondrite – a stony meteorite made of reprocessed materials.

Achromatic lens – a lens designed to correct chromatic aberration.

Acid rain – any precipitation that is unusually acidic.

Active galactic nucleus (AGN) – a compact region at the center of a galaxy that has a much higher than normal luminosity over some portion of the electromagnetic spectrum with characteristics indicating that the luminosity is not produced by stars.

Active optics – a procedure designed to keep the primary mirror in focus, regardless of changes in the weather or mechanical stresses.

Actuary – a business professional who deals with the measurement and management of risk and uncertainty researching the possibilities and probabilities involving their future operations.

Adaptive optics – technical procedures meant to correct the distortion of the image on a telescope caused by turbulence in the Earth's atmosphere.

Adiabaticity – the condition of being an adiabatic process (i.e. one that occurs without transfer of heat or mass of substances between a thermodynamic system and its surroundings).

Advanced computer simulations – computer software designed to simulate astrophysical phenomenon.

Aether – a supposed medium contemplated by Aristotle and Newton, which was theoretically the composition of stars and planets.

Air resistance – the resistance encountered by a moving object from air molecules.

Airglow – a faint emission of light by a planetary atmosphere. In the case of Earth's atmosphere, this optical phenomenon causes the night sky to never be completely dark, even after the effects of starlight and diffused Sunlight from the far side are removed.

Airy disk – the resulting diffraction pattern when imaging a star.

Albedo – the fraction of incident light that is reflected by a surface. Something that appears white reflects most of the light that hits it and has a high albedo, while something that looks dark absorbs most of the light that hits it, indicating a low albedo.

Alchemy – the medieval forerunner of chemistry, concerned with the transmutation of matter, in particular with attempts to convert base metals into gold or find a universal elixir.

Alkali elements – any of the six chemical elements that make up Group 1 (Ia) of the periodic table — namely, lithium (Li), sodium (Na), potassium (K), rubidium (Rb), cesium (Cs), and francium (Fr). The alkali metals are so called because reaction with water forms alkalies (i.e., strong bases capable of neutralizing acids).

Allele – a viable DNA (deoxyribonucleic acid) coding that occupies a given locus (position) on a chromosome.

Almagest – an astronomy encyclopedia published by Ptolemy in 0140 A.D.

Almanac – a yearbook predicting the position of all celestial bodies at all times.

Alpha decay – the sudden emission of alpha particles by a radioactive nucleus.

Alpha particle – a subatomic particle consisting of two protons and two neutrons. (i.e. a helium nucleus)

Alpha ray – a beam of alpha particles.

Altazimuth – a telescope mount which allows free movement of the apparatus in both vertical and horizontal axes.

Alternative histories – a genre of speculative fiction consisting of stories in which one or more historical events occur differently.

Amplitude – the maximum extent of a vibration or oscillation, measured from the position of equilibrium.

Analemma – a graphic representation predicting the position of the Sun at any time during the year.

Andromeda galaxy – a nearby large spiral galaxy located 2.5 million light years from the Milky Way.

Andromedids – a meteor shower which crossed the Earth's orbit in the 19th century.

Angstrom – a unit of length equal to 1×10^{-10} metres and often used in astronomy to measure wavelengths of light.

Angular diameter –the size of a distant object in units of an angle. Distance objects have smaller angular sizes than nearby objects.

Angular momentum – The angular momentum of a rigid object is defined as the product of the moment of inertia and the angular velocity. It is analogous to linear momentum and is subject to the fundamental constraints of the conservation of angular momentum principle if there is no external torque on the object.

Anisotropy – the quality of exhibiting properties with different values when measured along axes in different directions.

Annihilation – the destruction following a collision between a particle and its anti–particle.

Annual parallax – the angle subtended at a celestial object by the radius of the Earth's orbit.

Annular eclipse – an eclipse where the Sun is incompletely covered by the moon.

Anti–matter – a class of subatomic particle with identical "normal" particles except with opposite charge and spin.

Anomalistic month – the time required for the Moon to complete one orbit around the Earth and measured from its perigee.

Anomalistic year – the time required for the Earth to complete one orbit around the Sun and measured from its perihelion.

Anomaly – in astronomy, originally the non-uniform (anomalous) apparent motions of the planets. In present usage, three kinds of anomaly are distinguished to describe the position in the orbit of a planet, a satellite, or a star (in a binary system) around the center of mass.

Apparent laws – the current laws of nature acting in our Universe.

Apparent magnitude – a measurement of a celestial object's brightness in logarithmic units.

Apparent solar time – local time pertinent to the Sun's movement.

Apparition – the period of weeks or months during which a celestial body is visible in the night sky.

Apsides – the two points on an orbit that are respectively closest to (periapsis) and farthest from (apapsis) the focus of the orbit.

Arc minute – unit of angular measurement (1/60 of a degree).

Arc second – unit of angular measurement (1/3600 of a degree).

Archaea – a newly discovered realm of microbial life.

Aristotelian physics – physics promoted by Aristotle in which he proposed that the Universe is composed of four classical elements, Earth, air, water and fire. He also held that the heavens are made of a special weightless and incorruptible fifth element that he called "aether".

Arrow of time – the separation between past and present.

Asteroid – small, atmosphere-less rocky objects orbiting the Sun.

Asteroid belt – a zone between Mars and Jupiter where a large number of asteroids orbit.

Astrolabe – an ancient astronomical instrument for taking the altitude of the Sun or stars and for the solution of other problems in astronomy and navigation.

Astrology – a pseudo-science relating star movements to Earth happenings.

Astrometry – the science of the precision measurement of a star's location in the sky.

Astronomical twilight – the periods when the Sun is between 12 degrees and 18 degrees below the horizon.

Astronomy – the scientific study of celestial objects such as planets, stars, and galaxies, by direct observations.

Astrophysics – the scientific study of the physical nature of celestial objects and observed phenomena.

Astrophotography – the use of photography in the astronomical discipline.

Asymmetry – the counterpart of symmetry.

Asymptotic freedom – a property of some gauge theories that causes bonds between particles to become asymptotically weaker as energy increases and distance decreases.

Atmosphere – the gravitationally bound envelope of gases around a celestial body.

Atmosphere refraction – the image position of a celestial body as seen from the Earth through the Earth's atmosphere.

Atmospheric pressure – the pressure exerted by an air column above ground level.

Atom – the smallest particle still retaining the characteristics of a chemical element.

Atomic nucleus – the positively charged core of the atom.

Atomic number – the number of protons in the nucleus of an atom.

AU – Astronomical unit – the average distance between Earth and the Sun, which is about 150 million kilometres. Astronomical units are usually used to measure distances within our Solar System.

Aurora – a sudden natural light display in the night sky of the Earth (and of planets with a strong magnetic field), predominantly seen in the high-latitude regions (around the Arctic and Antarctic).

Avoidance – the tendency of the galaxies to elude the Milky Way.

Axion – a hypothetical subatomic particle postulated to account for the rarity of processes which break charge–parity symmetry.

Axion field – a group of axions acting in the same area.

Axis – a hypothetical line through the center of a celestial body joining its south and North Pole.

Azimuth – the angle measured westward from north on a vertical circle on a horizontal plane running through a celestial object.

B

Baryon – a composite subatomic particle made up of three quarks. The most stable baryons are protons and neutrons, so most building blocks of matter are baryons.

B mesons – exotic particles composed of a bottom antiquark and either an up (B^+), down (B^0), strange (B^0_s) or charm (B^+_c) quark.

B star – a class of hot stars with main sequence temperatures ranging from 10,600-25,000 Kelvin and exhibiting strong neutral helium absorption and some hydrogen lines.

Baily's beads – a brief event occurring before and after a total solar eclipse, a string of lights appearing at the edge of the Sun, at the very beginning or end of the eclipse.

Balmer series – the name given to a series of spectral emission lines of the hydrogen atom that result from electron transitions from higher levels down to the energy level with principal quantum number 2.

Barlow lens – a lens located between the eyepiece and the lens of a telescope in order to improve the quality of the image.

Barred spiral galaxy – a spiral galaxy with a band of bright stars emerging from the center and running across the middle of the galaxy.

Barycenter – the centre of mass celestial bodies orbiting around.

Baryogenesis – the physical process thought to give rise to the matter-antimatter asymmetry in the early universe.

Baryon number – conserved quantum number of all physical particles that have quark numbers divisible by 3.

BD – Bonner Durchmusterung is the comprehensive astrometric star catalogue of the whole sky, compiled by the Bonn Observatory from 1859 to 1903.

Bose-Einstein Condensate (BEC) – a state of matter in which separate atoms or subatomic particles, cooled to near absolute zero (0 K, − 273.15 °C, or − 459.67 °F; K = kelvin), coalesce into a single quantum mechanical entity — that is, one that can be described by a wave function — on a near macroscopic scale.

Bell's inequality – a discrepancy in the correlation of the quantum spins of entangled pairs of particles and the local hidden variables theory.

Bell's theorem – an important philosophical and mathematical statement in the theory of quantum mechanics. It showed that a category of physical theories called local hidden variables theory could not account for the degree of correlations between the spins of entangled electrons predicted by quantum theory.

Beryllium barrier – Prevention of the formation of elements heavier than helium in the early Universe due to the instability of the necessary element beryllium-8.

Beta decay – radioactive decay by emission of a beta particle.

Beta particle – a high–energy, high–speed electron or positron emitted by the radioactive decay of an atomic nucleus during the process of beta decay.

Beta ray – a beam of beta particles.

BeV – one billion electron volts.

Big bang – a theory of the beginning of the current Universe.

Big bang model (standard) – a model that proposes that the actual Universe started as a tiny bulk with an immense density and it has been expanding ever since.

Big bang Universe – an evolving Universe that had a singular origin in time.

Big crunch – a theory of the inverse of the big bang, assuming that at some point the Universe will stop expanding and that it will subsequently collapse.

Big rip – a theory that, due to the exceptionally high rate of expansion of the Universe, its components will eventually be torn apart.

Billion – giga (x 10^9).

Binary pulsar – a binary star system containing a pulsar and another binary companion.

Binary star – two stars orbiting together around the same centre of mass.

Binoculars – two small telescopes joined by a hinge aligned to point in the same direction, allowing the viewer to use both eyes.

Biodiversity – different forms of life in an ecosystem.

Biomarker – spectral features in a planet's atmosphere produced by life forms on its surface such as oxygen in the Earth's atmosphere produced by photosynthesis.

Bionic convergence – a futuristic idea that in the future humans may be implanted with electronic chips in the brain that will increase their intellectual potential.

Bipolar flow – star material flowing in opposite directions.

Bit – the smallest unit of data in a computing system.

BL Lacertae (also BL Lac) object – a particular bright and variable AGN

Black body – a celestial body that absorbs all incident electromagnetic radiation, regardless of frequency or angle of incidence.

Black body curve – the spectrum of a blackbody, showing that it is continuous (it gives off some light at all wavelengths) and that it has a peak at a specific wavelength that depends on the black body's temperature.

Black body radiation – the thermal electromagnetic radiation emitted by a black body.

Black drop effect – an optical phenomenon visible during a transit of Venus and, to a lesser extent, a transit of Mercury.

Black dwarf – a theoretical stellar remnant resulting from a cold white dwarf that has radiated all of its internal energy.

Black Eye Galaxy– a galaxy with a spectacular dark band of absorbing dust in front of the galaxy's bright nucleus, giving rise to its nicknames of the "Black Eye" or "Evil Eye" galaxy.

Black hole – a region of space with gravity so strong that even light can't escape.

Black hole complementarity – Bohr's principle of complementarity applied to black holes.

Black hole information paradox – the concept that any object which falls into a black hole gets destroyed.

Black hole spectrum – the spectrum emitted by a black hole.

Blazar – an active galactic nucleus (AGN) with a relativistic jet (a jet composed of ionized matter traveling at nearly the speed of light) directed very nearly towards the Earth.

Blink comparator – a microscope that optically superimposes two photographic plates, "blinking" between them so rapidly that the two plates look as one. This allows the researcher to compare the plates and look for any discrepancies between them.

Blue moon – either the third Full Moon of an astronomical season with four Full Moons or ~~as~~ the second Full Moon in a calendar month.

Blue straggler – a star which stays on the main sequence (the normal, hydrogen-burning phase of a star's lifetime) longer than it was expected to.

Blueshift – the decrease in wavelength (or corresponding increase in frequency) of an electromagnetic wave as it propagates towards an observer. The reverse effects is referred to as redshift.

Bode's Galaxy – a galaxy named after Johann Elert Bode, who discovered this galaxy in 1774.

Bode's law – a hypothesis that multiple planets in an orbital system should exist at approximately twice the distance for the central object as the one before.

Bohemian mechanics – a form of basic quantum theory which states that quantum randomness is not particular to particles.

Bok globules – very compact, isolated molecular clouds named for the astronomer Bart Bok, who first suggested they may be the precursors to protostars.

Bolide – A bolide is a meteor that is brighter than the planet Venus (i.e. a fireball), and which explodes or breaks up (usually accompanied by a sonic boom) before reaching the ground.

Bolometer – a device used to measure incoming power from a radiation source by the degree of heating induced by said radiation.

Bolometric magnitude – a measure of the total radiation of a star emitted across all wavelengths of the electromagnetic spectrum. The absolute bolometric magnitude (abm) is the bolometric magnitude the star would have if it was placed at a distance of 10 parsecs from the Earth.

Bootstrap paradox – a paradox of time travel where an object creates itself.

Bosonic string theory – an older version of the String theory developed in the 1960s that only contains bosons in the spectrum.

Bosons – a class of force carrying subatomic particles with integer spin and are described by Bose-Einstein statistics. An example is a photon.

Bottom approach – a concept that there is only one version of the Universe and that the actual Universe is the final product of a long evolution from the big bang.

Bottom quark – a massive third generation quark with negative one third of the elementary charge.

Bounce – a method for creating a supernova, by producing a shock wave that rips the outer part of the core of the star.

Boundary conditions – the demarcation line of the extent of an equation.

Boundary theory – the mathematical version describing everything inside a certain region.

Bow shock – the region around a planet were the magnetosphere prevails over the solar wind.

Brane – a celestial object which is extended in one or more dimensions.

Braneworld – a world that is a brane, with its own limited set of dimensions, in any of several brane cosmology theories.

Breccia – a celestial body composed of rock fragments accumulated by the force of gravity appearing as a single rock.

Bremsstrahlung (brake radiation) – a form of radiation emitted by accelerating or decelerating a charged particle such as an electron.

Broken symmetry – is a phenomenon in which (infinitesimally) small fluctuations acting on a system crossing a critical point decide the system's fate, by determining which branch of a bifurcation is taken.

Brown dwarf – a small celestial object which cannot reach the necessary temperature in order to have a nuclear fusion reaction in its core and must therefore continue to radiate it's own internal thermal energy as infrared radiation. Also known as failed star.

Brownian motion – the random motion of particles suspended in a fluid.

Brute fact – a fact which simply is, without the need of an explanation.

Bubble – a quantum process whereby a field trapped in a false vacuum by an energy barrier escapes, generating a spherical volume of true vacuum.

Bucky space – a name given by George Musser to causal dynamical triangulations.

Butterfly diagram – a graphical chart pointing out the locations of Sunspots over a period of 11 years.

Calaby–yau manifold – a method in string theory for compressing the spatial dimensions.

Calendar – a graphical representation of the time in a year, divided into days, months, etc.

Caltech – the California Institute of Technology in Pasadena.

Carbon cycle – a natural process that moves carbon compounds from the organic material, to the Earth's oceans, interior, and atmosphere.

Carbon reaction – chemical reactions involving the element carbon.

Carbon star – a red giant star whose atmosphere contains more carbon than oxygen

Carbonaceouscondrite – a chondrite meteorite containing more than the normal quantity of carbon.

Carbon-nitrogen-oxygen cycle (CNO cycle) - a nuclear fusion reaction ongoing in some stars that fuses hydrogen into helium using carbon, nitrogen, and oxygen as catalysts.

Cartwheel Galaxy – a galaxy so named because of its appearance similar to that of a spoked cartwheel.

Casimir effect – the phenomenon of two near flat and electrically neutral metal plates attracting each other in a vacuum.

Cassegrain telescope – a kind of telescope which redirects the light from the primary mirror to the secondary mirror through a special orifice.

Cassini division – a large space between the rings of Saturn.

Cataclysmic variable – a binary star system consisting of a white dwarf and a normal star companion that transfers some of its mass onto the white dwarf thus creating brightness variability.

Catadioptric – an optical system in which refraction and reflection are combined, usually via lenses (dioptrics) and curved mirrors (catoptrics).

Catastrophism – the idea that many of Earth's crustal features (strata layers, erosion, polystrate fossils, etc) formed as a result of past cataclysmic activity.

Causal contact – state by which particles are reached from a common point in the past, no matter how far apart they are physically

Causal dynamical triangulation – a quantum theory of gravity that divides the space in virtual triangles.

Causality – the concept that every new state is a~~ the sequel to a previous condition.

CDMS – cryogenic dark matter search.

Celestial – positioned in or relating to the sky, or outer space as observed in astronomy.

Celestial equator – the projection of the Earth's equator onto the sky.

Celestial mechanics – a branch of astronomy studying the evolution of the motion of celestial objects.

Celestial meridian – an abstract circle cutting through the observer's horizon from north to south and zenith and nadir.

Celestial sphere – an imagined sphere around the Earth with the celestial objects projected upon it.

Celsius – a scale for measuring temperature.

Centenarian – a human who has lived 100 years or longer.

Center of mass – a single point at which the whole mass of the body or system is imagined to be concentrated with all the applied forces acting on that point.

Central meridian – an imaginary line connecting the north and south poles.

Centrifugal force – a force, arising from the body's inertia, which appears to act on a body moving in a circular path and is directed away from the center around which the body is moving.

Centripetal force – a force which acts on a body moving in a circular path and is directed towards the centre around which the body is moving.

Cepheid variable – a pulsating variable star whose period is related to its absolute magnitude.

CERN – Conseil Européen pour la Recherche Nucléaire, the laboratory in Switzerland hosting the Large Hadron Collider.

Chance – an observation of nature by which prediction cannot be made with precision, just placed in the realm of probabilities.

Chandrasekhar limit – the maximum mass of a white dwarf starequal to 1.4 Solar masses.

Chandrayaan – lunar probe launched in October 2008 by the Indian Space Research Organization.

Chaotic inflation – the concept that the Universe expanded from the big bang and is subject to rapidly accelerating expansion.

Charge–coupled device (CCD) – an electronic device used to capture imaging information in digital cameras.

Charm quark – a second generation massive quark with positive two thirds of the elementary charge.

Chemical bonds – bonds that facilitate the outer electrons of atoms combining into new molecules.

Chicxulub impact – a large impact crater on the Yucatan peninsula from about 65 million years ago.

Chiral molecules – molecules with right and left hand structure.

Chondrite – stony (non–metallic) meteorites that have not been modified due to melting or differentiation of the parent body.

Chondrule – a spheroidal mineral grain present in large numbers in some stony meteorites.

Chromatic aberration – a defect in lenses, affecting the image in a telescopedue to the refraction of different wavelengths of light.

Chromosphere – a layer of the Sun's atmosphere between the photosphere and the corona.

Chronometer – a precise timepiece.

Circle – an ellipse with only one focus at its centre.

Circumpolar stars – stars located close to the either celestial pole that remain above the observer's horizon as the Earth rotates.

Classical – a term in physics referring to a pre–quantum theory.

Classical mechanics – the Newtonian way of describing the motions of physical systems in the Universe.

Classical physics – Newtonian physics from the pre-quantum and pre-relativity eras.

Climate change – variations in global climate over time.

Clipper chip – a cryptographic device purportedly intended to protect private communications while at the same time permitting government agents to obtain the "keys" upon presentation of what has been vaguely characterized as "legal authorization."

Clonal colony – a group of genetically identical individuals.

Closed string – a string without end.

Closed Universe – a theoretical "ideal" Universe with enough matter to stop its expansion.

Cloud chamber – a glass recipient containing a vapor where particles leave behind tracks of water droplets which can be photographed.

Cluster of galaxies – a gravitationally bound group of galaxies.

COBE – the Cosmic Background Explorer, a satellite dedicated to measuring the cosmic microwave background radiation of the universe.

Coelostat – an instrument with a rotating mirror that continuously reflects the light from the same area of sky, used for monitoring the path of a celestial object.

Coherent – a synchronized beam vibration.

Coincidence scandal – the cosmological problem posed by the fact that the energy density of matter is the same as the energy density of the vacuum, whereas in theory they should not coincide.

Cold dark matter – a hypothetical type of dark matter with characteristically low velocities.

Collapse of the wave function – the change from a system that can be seen as having many possible quantum states (Dirac's principle of superposition) to its randomly being found in only one of those possible states.

Collider – an apparatus that accelerate particles in order to collide them with stationary particles, so as to examine the accelerated particles.

Collimation – a method to narrow a beam of radiation often used for the alignment of elements in an optical system.

Colour – an attribute of a wave of light in space related to its wavelength.

Colour index – the difference between two apparent magnitudes of an object and indicative of its temperature.

Colure – either of two great circles of the celestial sphere intersecting each other at right angles at the celestial poles.

Coma (cometary) – a gaseous envelope surrounding a comet's nucleus.

Coma (optics) – a deformation of the image, due to a defect in the optical system.

Comet – a small icy-rocky object originating from the outer Solar system that produces a coma and cometary tail when traveling through the inner Solar system as it is heated by the Sun.

Commensurable – a common factor for two numbers.

Commutativity – the ability of two variables to be multiplied in any order.

Comoving viewpoint – a viewpoint that expands with the Universe.

Compactification – a process were large spatial dimensions are shrinking, leaving some space free.

Comparator – an apparatus which permits the viewing of two photographs of the same object at different times by quickly alternating their view.

Complementary – a principle that the wave and particle aspects of matter are complementary but exclusive, in the sense that both cannot be displayed simultaneously.

Complex molecules – a collection of atoms bound together into large molecules and acting as a whole.

Complex number – concrete numbers having a real part from arithmetic as well as an imaginary part.

Compton effect – the scattering of photons by a charged particle such as an electron.

Computational astrophysics – a discipline of astrophysics that relies on computing power to simulate various astrophysical phenomena within the Universe.

Concordance cosmology – a theory which attempts to unify all the dynamics in the Universe.

Concordance model – a cosmological graphic representation of the matter spread in the Universe.

Confinement – a state in which quarks that are held together by the strong nuclear force e.g. single quarks, are confined by interaction into combinations such as baryons and mesons.

Conjugate variables – a pair of dynamic variables that are related through the uncertainty principle.

Conjunction – two different bodies in our galaxy so aligned that they appear from the Earth as being in the same position.

Conservation law – a law that states that a physical quantity cannot be created or destroyed, even if it goes through a physical process.

Conservation of energy – the principle that energy cannot be destroyed nor created, just transformed between different types of energy.

Conserved quantity – the quantity that is conserved under a conservation law (i.e. energy under the conservation of energy)

Constellation – a group of stars which seems to outline an Earthly figure on the celestial sphere.

Continuous spectrum – the intensity of light from a source over wavelengths in the absence of absorption or emission lines.

Convective zone – a zone of the Sun's core where energy is transported by the convective motion of the gases.

Convergence – the tendency of a system to move towards a common union.

Co–orbital satellite – a satellite orbiting a planet at the same distance as another satellite.

Coordinates – figures that establish a point in space and time.

Copenhagen interpretation – a classical interpretation of quantum mechanics as developed by a core group of quantum physics pioneers centered around Niels Bohr's Copenhagen Institute through the 1920's.

Copernicanism – movement relating to Copernicus or the belief that the Earth rotates daily on its axis and the planets revolve in orbits around the Sun.

Core of the Sun – the centre of the Sun where nuclear fusion takes place.

Coriolis force – an inertial or fictitious force that seems to act on objects that are in motion within a frame of reference that rotates with respect to an inertial frame.

Corona – a peripheral layer of plasma that surrounds the Sun and constitutes the outermost layer of its atmosphere.

Coronal hole – the coolest region of the Sun's atmosphere.

Coronal loop – a loop made by the Sun's magnetic field from the Sun's surface through its corona and back.

Coronal mass ejection (CME) – the ejection into space of huge quantities of plasma from the Sun's corona.

Coronium – an element which is part of the corona.

Coronograph – a special device which obstructs the Sun's light in order to allow the observation of Sunspots and the corona.

Corpuscles – a Newtonian term for the imaginary particles of light.

Correspondence principle – the principle that states that predictions of quantum mechanics must match the predictions of classical physics in the physical situations that classical physics is intended to describe, and does describe very accurately.

Cosmic abundance – the relative amounts of chemical elements that exist in the Universe.

Cosmic censorship conjecture – theory stating that bare singularities are not workable.

Cosmic coincidence – an incidental occurrence whereby energy was prevalent in the Universe while the stars and galaxies were taking shape.

Cosmic evolution – changes in the Cosmos as it evolves in time.

Cosmic strings – strings stretched across the Universe.

Cosmological scandal – the question of why everything coincided in such a way that life on Earth could come into existence.

Cosmologically speaking – according to cosmological parameters.

Critical density – the density value at which the Universe is at balance, and expansion is stopped. This value is estimated as $(1-3) \times 10^{-26}$ kg/m³ and is calculated by taking the matter–energy density of the Universe and dividing it by the matter–energy density of the Universe that is required to achieve that balance.

Cosmic matter density – the average number of fermions per unit volume in the entire Universe.

Cosmic microwave background (CMB) – the relic microwave radiation from the early Universe that fills the entire sky.

Cosmic rays – high energy particles from space running through the Earth's atmosphere.

Cosmogony – the study of the Universe, its formation, and of the objects belonging to it.

Cosmological constant – the vacuum energy of dark energy.

Cosmological constant problem – the discrepancy between the amount of vacuum energy indicated and the amount of vacuum energy expected according to calculations.

Cosmological principle – the proposition that on a large scale, the Universe seems isotropic and homogeneous.

Cosmological redshift – the redshift of photons resulting from the expansion of the Universe.

Cosmology – a comprehensive study of the Universe, including its origin, configuration, and evolution.

Cosmos – the known Universe.

Cosmos Redshift 7 – the brightest of distant galaxies ($z > 6$) containing some of the Universe's earliest stars.

Coude – an alignment of mirrors in a telescope meant to adjust the focus.

Coulomb barrier – the electromagnetic field around charged particles that repels other similarly charged particles.

Counterfactual – relating to something that is not the case.

CP symmetry – charge conjugation parity symmetry, the conservation of electric charge and polarity.

CP violation – the incident of particles not respecting the CP symmetry.

Crater – a circular structure on the surface of a celestial body in the solar system caused by an impact with another celestial object.

Creation myth – a popular account of the creation of the Universe.

Creationism – belief that the Universe was created by God.

Cretaceous–tertiary boundary – the geological division between the cretaceous and tertiary periods of the Earth's history.

Critical density – the amount of density necessary in order to have everything flat in space–time.

Crust – the outermost layer of the Earth enclosing the mantle and the core.

Cryonics – a preservation of living things like humans through a method of low temperature.

Cubic inch, cubic centimetre, cubic foot, etc – different units to measure volume.

Culmination – the crossing of a celestial body through the observer's meridian.

Current distance – the distance to another celestial body at a point in time (given that everything in the Universe is moving).

Curvature – the degree of bending of space–time.

Curvature of space – according to Einstein's general theory of relativity, the distortion of space–time by matter.

CXO – Chandra X–ray Observatory.

Cyberspace – a hypothetical world of computers connected by a network like the internet.

Cyborg – a hypothetical experiment of providing the human body with technological devices in order to enhance human bodies with extra capabilities.

Cyclic model – any of several cosmological models in which the Universe follows infinite, or indefinite, self-sustaining cycles.

D

Dama – the Gran Sasso laboratories in Italy

Dark adaptation – a property of the human eye allowing it to adjust its capacity of seeing from high illumination to darkness.

Dark energy – an unknown force which acts as an anti–gravity pressure and is responsible for the acceleration of the expansion of Universe.

Dark energy era – the point in time (9 billion years ago) when dark energy began to dominate the cosmic density over matter.

Dark matter – invisible matter which is presumed to make up 90% of the matter in the Universe but whose constituent particles remain unknown.

Dark nebula – a type of interstellar cloud that is so dense that it obscures the light from objects behind it, such as background stars and emission or reflection nebulae.

Dark star – a star that has a gravitational pull strong enough to trap light under Newtonian gravity.

Darwinism – theory that species arise and survive due to a natural selection of the fittest individual.

Day – a period of time in which the Earth turns once on its axis.

D–brane – a layer in space–time where the basic strings may end.

Debris disk – a disk orbiting a star that is primarily composed of cold, low density rocks and ices.

De broglie wavelength – a wavelength, which is manifested in all the particles in quantum mechanics, according to wave–particle duality, which determines the probability density of finding the object at a given point of the configuration space.

De revolutionibusorbiumcoelestium – the seminal work on the heliocentric theory of the Renaissance astronomer Nicolaus Copernicus supporting the idea that the Sun is the centre of the Universe.

Decay – a natural breakdown of a particle into its components.

Deceleration parameter – the number determining the rate of retardation of a body's current speed.

Decibel (dB) – a unit used to measure the intensity of a sound or the power level of an electrical signal by comparing it with a given level on a logarithmic scale.

Declination – the angular position of a celestial object, relative to the celestial equator.

Decoherence – an action through which objects lose their characteristics becoming ordinary matter.

Decouple – the act of separation such that two components are no longer in contact.

Decoupling era – the period after the big bang when conditions allowed the formation of atoms via the decoupling of photons and electrons.

Deduction – a way of reasoning, starting with the available premises.

Deep field – a portion of the sky selected to be photographed by the Hubble telescope in a multitude of images.

Deep sky – the Universe beyond our solar system.

Degeneracy pressure – a physical pressure exerted by electrons and neutrons when they are restricted in a small area.

Degenerate matter – matter that is internally supported by degeneracy pressure.

Degree – a unit of measurement of an angle subtended on the sky.

Degrees of freedom – directions of movement of a system (if it can move).

Density fluctuation – slight variations in the density of matter in some regions of the Universe.

Density parameter – the ratio of the density of the Universe relative to the critical density.

Density – the mass per unit volume of a body.

De sitter space – the analog in Minkowski space, or spacetime, of a sphere in ordinary Euclidean space.

Descending node – the crossing point of an orbit with the equator or the ecliptic.

Describing the Universe – presenting the Universe as seen through the actual apparatus of observation.

Detector – an apparatus which registers the existence of subatomic particles.

Determinism – the classical principle proclaiming that the future is determined by the present.

Deuterium – an isotope of hydrogen with a nucleus composed of one neutron and one proton.

Deuterium bottleneck – a delay in proceedings until the deuterium nucleus is ready for fusion in order to produce helium.

Diagonal – a type of telescope eyepiece that offers a perpendicular viewing angle for the sake of comfort.

Diamond–ring effect – an unusual apparition during a total solar eclipse, when an intense brightness of light seen through a weak corona at the edge of the Sun looks like a diamond ring.

Dichotomy – the appearance of Venus or Mercury half illuminated by the Sun.

Differential rotation – the phenomenon in a non–solid body where the parts rotate at different speeds.

Differentiation – the process by which a planetary body develops different layers.

Diffraction – a ray of light directed sideways when it passes through a narrow passage or around an edge.

Diffraction grating – a plane bearing a multitude of parallel slits at the same distance from each otherfor the purpose of separating light into colours to produce a spectrum.

Diffuse nebula – an illuminated gas cloud in space.

Dilation – the expansion of space or time relative to a fixed point.

Dimension – a geometric parameter describing the minimum number of coordinates required to specify a point.

Dynamical variables – the characteristics of the state of a particle (momentum, potential and kinetic energy).

Dipole – a pair of magnetic charges, of equal magnitude and opposite alignment, separated by some distance.

Dirac equation – an equation which unifies quantum mechanics and special relativity in order to arrive at a mathematical description of the electron.

Direct detection – the detection of an object by seeing it rather than inferring its presence based on its interaction with another directly detected object.

Direct motion – direction of motion in space that is similar to that of the Earth.

Disconnection event – the detachment and eventual recombining of a comet ion tail.

Discrete – a type of quantity that can only exhibit particular values unlike quantities that are continuous.

Disk galaxy – a designation given to all galaxies in the form of a disk.

Dispersion – the variation of the index of refraction of a transparent substance, as glass, with the wavelength of light, with the index of refraction increasing as the wavelength decreases.

Dissipationless collapse – the gravitational collapse of a system into a low density system without losing energy.

Dissipative collapse – the gravitational collapse of a system into a high density system in which energy is lost by radiation.

Distance ladder – a class of techniques used to derive distances to celestial objects. Different techniques are useful over different magnitudes of distances.

Distance modulus – the discrepancy between the apparent and the absolute magnitude of a star

Diurnal motion – the apparent daily motion of the celestial bodies across the sky from east to west.

DNA – the molecule that contains the genetic information necessary for life (deoxyribonucleic acid).

Dobsonian telescope – an altazimuth–mounted Newtonian telescope design popularized by John Dobson in 1965.

Dollar matrix – proposed by Stephen Hawking, a matrix describing the transformation from particles to black hole to Hawking radiation. It could not be described with an S–matrix, so Hawking proposed a "not–S–matrix", for which he used the dollar sign, and which therefore was also called the "dollar matrix".

Doppler effect – the apparent increase or decrease in light's wavelength as it travel towards or away from the observer respectively.

Double star – two stars that present the observer with an image of being close together.

Doublet – two pieces of a lens installed close together or with a slight separation in between.

Doubly special relativity – converted version of Einstein's special theory of relativity, having the speed of light and plank scale as universal limits.

Down quark – a first generation quark with negative one third of the elementary charge and together with top quarks make up protons and neutrons.

Draconic month – a period of time between two consecutive passages of the Moon through the same orbital node.

Drake equation – a sequence of numerical factors conducing an estimate of the probable existence of actual intelligent life in our galaxy (the Milky Way).

Drive – a way of compensating for the rotation of the Earth, by displacing a telescope according to the Earth's movement.

Duality – concept that two distinct connected situations may be the same if they end up with the same results.

Dumb hole – a drain hole were the velocity of the flow exceeds the speed of sound.

Dust cloud – gas clouds in interstellar space, which contain dust molecules in the stage of formation.

Dust devil – a small storm tornado – on Mars.

Dust tail – a comet tail pressured back by the solar wind.

Dwarf galaxy – a small galaxy with low luminosity.

Dwarf nova – an irregular variable star with light curves that resemble that of a novae.

Dwarf planet – a near-planetary mass object that does not meet the criteria of being classified as a planet owing to the quantity of interplanetary debris in the immediate environment of its orbit.

Dwarf star – a class of star describing stars on the main sequence regardless of their mass or temperature.

Dynamical parallax – an equation comprising the mass, the size of the orbit and the period of a star revolution, revealing the distance of a binary star.

Dynamics – the study in physics of the forces, motion, and balance of all the moving parts in a system.

Dynamo – a way of creating a magnetic field in the core of stars and planets by way of rotation and convection of a conducting fluid.

Early–type star – a hot star of spectral type O, B, or A.

Earth – the third major planet from the Sun in the solar system.

Earth–crossing asteroid – asteroid which may present a remote possibility of hitting the Earth.

Earthgrazer – a very bright meteor that enters Earth's atmosphere and leaves again.

Earthshine – sunlight reflected from the Earth.

Eccentric – not placed centrally or not having its axis or other part placed centrally.

Eccentric circles – circles that do not share the same center although the centers of each circle are all contained within at least one of the circles.

Eccentricity – a measure of the how elliptical an orbit is. Bound orbits with increasing eccentricity from zero to one increase in ellipticity are correspondingly less circular.

Echelle grating – a diffraction grating in which the grooves are quite widely spaced and have a zigzag or step–like cross–sectional profile.

Echelle spectrograph – a spectrograph for measuring the spectra of celestial objects that employs an Echelle grating.

Eclipse – the total or partial blocking of a celestial body by another relative to an observer.

Eclipse year – the interval of 346.62 sidereal days between two successive conjunctions of the Sun with the same node of the moon's orbit.

Eclipsing binary – a binary star whose brightness varies periodically as the two components pass one in front of the other.

Ecliptic limits – the greatest angular distance between the Sun and the moon.

Ecosphere – the envelope of space surrounding a celestial body.

Ecosystem – a community of living organisms interacting with one another and their environment.

Eddington limit – the greatest luminosity an object can have without being torn apart by radiation pressure.

Effective temperature – the hypothetical temperature of a star, considering that its output will be in the form of black body radiation.

Einstein's equation – the cornerstone of his general theory of relativity, they describe how the distortions of space–time are connected with the properties (mass, energy, pressure...) of whatever matter is present.

Einstein's ring – a circular image from a distant source produced by a gravitational lens.

Einstein–Rosen bridge – a minor tunnel of space–time linking two black holes.

Ejecta – material ejected from two bodies after a collision.

Ekpyrosis – a collision of two branes resulting in a new flat Universe full of matter and radiation.

Electric charge – the physical property of matter that causes it to experience a force when placed in an electromagnetic field.

Electric current – an amount of electric charge moving through a conductor or electric space per unit time.

Electric field – the force field created by the presence of an electric charge. Analogous to the gravitational field created by a massive object.

Electromagnetic radiation – a form of energy that is propagated by photons.

Electromagnetic spectrum – the full extent of frequencies exhibited by electromagnetic radiation. Examples include radio, infrared, optical, and X-ray radiation.

Electromagnetic waves – electric and magnetic fields that propagate at the speed of light and oscillate with a characteristic wavelength and frequency.

Electromagnetism – the description of electric and magnetic fields as one unified force.

Electron – a negatively charged subatomic particle that together with atomic nuclei form neutral atoms.

Electron neutrino – a subatomic lepton elementary particle which has no net electric charge.

Electron shells – classical regions surrounding an atomic nucleus hosting electrons and each with a unique energy state.

Electron volt – the basic unit of electrical energy.

Electronuclear force – a combined force of the electromagnetic, the strong and weak nuclear force.

Electroweak theory – theory particularizing quantum electrodynamics.

Electroweak unification energy – energy level at which these two forces join, becoming one single force.

Element, chemical – a basic material which composes all matter in the Universe.

Elementary particle – fundamental object without internal structure.

Ellipse – an elongated circle.

Elliptical galaxy – a class of galaxy with an elliptical morphology (unlike disk galaxies).

Elongation – the difference in degrees of celestial longitude between two spatial objects.

Embedding diagram – a description of space–time at a certain point.

Emergence – procedure of a system by which it acquires new components.

Emersion – the emergence of a celestial body obscured by another celestial body during an eclipse.

Emission distance – the distance from a celestial body to where its radiation could be received.

Emission lines – spectral lines resulting from electrons in a hot gas dropping down to lower energy states and consequently emitting a photon at a fixed wavelength.

Emission nebula – a nebula formed from ionized gases and exhibiting strong emission lines.

Emission spectrum – a spectrum containing emission lines.

Endothermic – a reaction where energy is absorbed over time.

Energy – ability to do work.

Energy barrier – a stack in the energy curve obstructing the evolution of a field towards the minimum.

Energy curve – a graphical representation of an energy value.

Energy density – energy per unit volume.

Entanglement – the operation by which two particles connect and remain connected no matter what external changes occur.

Entropy – a thermodynamic quantity representing the unavailability of a system's thermal energy for conversion into mechanical work, often interpreted as the degree of disorder or randomness in the system.

Entropy of the Universe – the degree of disorder ensuing in the Universe.

Environmental selection – the fact that humans consider only the environments pertinent to their existence.

Ephemeris – the positions of celestial objects at a given time or times.

Epoch – a point in time used as a reference point for some time–varying astronomical quantity, such as the celestial coordinates or elliptical orbital elements of a celestial body, because these are subject to perturbations and vary with time.

Epicycle and deferent – in both Hipparchian and Ptolemaic systems, the assumed movements of the planets in a small circle called an epicycle, which in turn moves along a larger circle called a deferent.

Epicycles – part of Ptolemy's model of the Universe, circles that were supposed to be the paths of the planets.

Equality – the point in time when the radiation plus the neutrino density equals the matter density.

Equant – a geometric model of the solar system in which the planets move at a uniform speed.

Equation of state parameter – formula that indicates the eventual deviation from the constant dark energy density parameter.

Equation of the centre – an aberration in the course of a body moving in an elliptical orbit.

Equation of time – the relation between the time shown on a clock and the time exhibited by a Sun dial.

Equatorial telescope – a telescope installed in such a way that it can control the polar as well as the declination axis.

Equinoctial colure – a circle passing through the north and south celestial poles and the equinoxes.

Equinox – point at which the Sun crosses the celestial equator twice during an Earth year.

Equivalence principle – theory that the acceleration of a body is linked to gravity.

Erfle eyepiece – an eyepiece of a telescope designed to widen the image rather than showing its depth.

Erg – a unit of energy.

E.S.A. – European Space Agency.

Escape velocity – the minimum velocity required by a body to be able to overcome Earth's gravity and escape into space.

E.S.O. – European Southern Observatory.

Eternal inflation – the concept that the Universe will expand forever.

Euclidean geometry – the geometrical rules for flat spaces proposed by Euclid in around 300 B.C.

Euclidean time – a mathematical system expressing time along imaginary axis.

Eukaryote – a cell containing genetic material in its nucleus.

EUVE – Extreme Ultraviolet Explorer.

Evaporation – a process by which a substance phase changes to a gaseous state from a liquid state.

Event – a moment in space–time

Event horizon – the region around a black hole inside of which light cannot escape.

Exclusion principle – the principle that two fermions cannot simultaneously occupy the same quantum state within a quantum system.

Exobiology – the study of life outside of the Earth. Also known as astrobiology.

Exothermic – chemical or nuclear reactions accompanied by the release of heat energy.

Extra dimension – a dimension yet to be detected but which acts besides the three known spatial dimensions.

Fabry–perrot interferometer – a special device for measuring close spectral lines.

Faculae – bright regions surrounding Sunspots in the Sun's photosphere.

Falling star – a meteor.

Family – the fundamental particles of the standard model.

Far–infrared – the region of the infrared range of the electromagnetic spectrum with the longest wavelengths.

Fermi's paradox – if there is extraterrestrial intelligence in the Universe, why has it not visited us?

Fermilab – the Fermi national accelerator laboratory in Batavia, Illinois.

Fermions – ordinary particles of matter with half integer spins.

Feynman diagram – a graphical representation of particle dynamics.

Field (astronomy) – the area of space against the solar disk.

Field (physics) – the region where a force operates.

Field of view – the area visible through an optical instrument.

Fifth force – a hypothetical fifth fundamental force, in addition to the four known fundamental forces, that is thought to be responsible for the effects of dark energy in the Universe.

Filament – a prominence in the solar system appearing as a silhouette.

Filar micrometer – a specialized eyepiece used in astronomical telescopes for astrometry measurements, in microscopes for specimen measurements, and in alignment and surveying telescopes for measuring angles and distances of nearby objects.

Filter – an optical device used to restrict incoming light to a specific region of the electromagnetic spectrum.

Finder – a smaller telescope used to locate objects in space to be observed by the main telescope.

Fine structure – the splitting of an energy level or spectral line into its components.

Fireball – an unusually bright meteor.

First law of thermodynamics – the law of conservation of energy.

First light – the first time a telescope or instrument conducts an astronomical observation.

First quarter – a moon phase in which half of the Moon's surface is illuminated.

Fission – a nuclear reaction that involves splitting of an atomic nucleus.

Flamsteed numbers – designations attributed to stars.

Flare – an explosion on the Sun's surface.

Flare star – a star which displays bursts of brilliant light sporadically.

Flash spectrum – the exposure of the solar chromospheres, visible during a solar eclipse.

Flat plane – Euclidean geometry presenting no curvature.

Flatness problem – a problem of fine tuning in producing a flat Universe because the Universe is flat only under certain conditions.

Flavor – refers to the various types of quarks and leptons.

Flocculi – minor particles creating a fluffy image in the Sun's chromospheres.

Flop transition – dynamic activity in the Calaby–Yau region of the sky, with moderate outcomes in string theory.

Focal length – the distance between the center of a lens and its focus.

Focal ratio – the ratio of the system's focal length to the diameter of the entrance pupil.

Following – tracing the movements of a celestial body.

Forbidden lines – a class of absorption or emission lines that are not normally allowed under a particular transition rule.

Force – action at a distance producing an observable change in the system being acted on by the force.

Fossils – remains of ancient living things.

Fourier transform – a mathematical technique used to determine the amplitude of sine waves involved in a function.

Fractal – a class of complex geometrical shapes that exhibit self–similarity at all scales and show an endless repeated geometry.

Fraunhoffer lines – dark absorption lines in the solar spectrum.

Free radical – an atom or molecule with an independent electron in its peripheral orbit.

Frequency – measure of the number of cycles per time of a wave.

F–star – a class of stars with main sequence temperatures ranging from 6000-7000 Kelvin and exhibiting strong absorption from hydrogen and ionized metals.

Fundamental charge – the charge carried by the electron and the proton. Also known as the elementary charge.

Fundamental stars – well-established stars with a steady orbit.

Fundamental strings – the basic strings that make up gravitons.

Fundamental tone – the lowest possible frequency produced by a musical instrument.

FUSE – Far Ultraviolet Spectroscopic Explorer.

Fusion – a nuclear reaction that forms heavy elements by combining multiple light elements.

Futurology – the scientific prediction of the future.

G

Gaia – the theory that the biosphere adjusts according to the process of evolution in order to preserve life.

Galactic cannibalism – the incorporation of a smaller galaxy into a bigger galaxy system.

Galactic disc – the component of a galaxy whose morphology resembles a disk.

Galactic dynamics – the study of the motions of objects (i.e. stars, gas, dark matter, etc.) within a galaxy.

Galactic halo – the spread of stars and stellar clusters around the galaxy.

Galaxy – a gravitationally bound collection of stars, gas, dust, and dark matter.

Galilean principle of gravity – the principle that uniform motion must have a reference point as there is no absolute standard of rest since the laws of mechanics are the same for all observers in uniform motion.

Galilean telescope – a refracting telescope having a plano-convex objective and a plano-concave eyepiece.

Galilean transformation – a change of reference frames for measurements in which the two reference frames are in uniform linear motion with respect to one another.

Gamma rays – ultra high energy electromagnetic radiation.

Gamma–ray light – the shortest wavelength and highest energy light in the electromagnetic spectrum.

Gamma–ray astronomy – the study of high–energy electromagnetic radiation from space.

Gamma–ray burst – an intense flash of gamma rays which may last from about a second to a few minutes.

Gas cloud – a diffuse collection of particles in interstellar space. It is composed of hydrogen, helium, and traces of the heavier elements.

Gauge boson – a force carrier, a bosonic particle that carries any of the fundamental interactions of nature, commonly called forces.

Gauge symmetry – a state whereby, under universal changes, the gauge symmetry is working, but where local changes prevail, the local symmetry applies.

Gauge theory – concept that some forces originate from broken symmetry.

Gegenshein – a faint, luminous patch on the ecliptic, visible usually in the tropics, on dark, clear nights, directly opposite the Sun.

Geminga – one of the most powerful gamma–ray sources in the sky, in the constellation Gemini.

Geminids – a meteor shower, occurring in December.

General relativity – theory proposed by Einstein that proclaims that the laws of science should be the same for all observers, no matter how they are moving. It explains the force of gravity in terms of a curvature of a four dimensional space–time.

Generation (particle physics) – a periodic group of matter particles from the standard model.

Genome – the DNA of an organism containing hereditary information.

Genotype – a set of genes in an organism responsible for a single trait, even if it is not fully expressed.

Geocentric – Earth-centred.

Geodesic – the shortest path between two points through space-time.

Geology – the scientific study of the Earth's past through evidence left in the Earth's remains like rocks, Earth layers, and fossils.

Geometry – a mathematical discipline dealing with the shapes, sizes, and positions of figures.

German mounting – a special type of mounting, for an equatorial telescope.

GeV – giga–electron volt.

Giant planet – a class of large planets primarily made up of gas.

Giant star – a star with substantially larger radius and luminosity than a main–sequence star of the same surface temperature. They lie above the main sequence on the Hertzsprung–Russell diagram and correspond to luminosity classes II and III.

Gibbous – a phase of the moon between the half and full Moon phases.

Giga byte – a unit of memory storage equal to 10^9 bytes.

GLAST – Fermi Gamma–ray Space Telescope, formerly the Gamma–ray Large Area Space Telescope.

Glitch – a burst of light during a pulsar's period.

Global symmetry – symmetry which acts the same way all over the Universe.

Global warming – the rising temperatures in the past century on our planet.

Globular cluster – an ancient cluster of old stars, evolving in our galaxy.

Glue ball – a hadron composed only of gluons with no quarks.

Gluon lattice – a force field that holds quarks together, generated by the strong nuclear force.

Gluon – a particle that conveys the strong nuclear force keeping quarks together to form protons and neutrons.

G.M.C. – Giant Molecular Cloud.

Gnome – a complete organism's information contained in its DNA, or, for some viruses, RNA.

GNOG – Global Network Oscillation Group.

Gould's belt – a zone of bright stars and gas in the galactic plane.

GPS – Global Positioning System: a precise satellite–based location system used for navigation.

GR – general relativity.

Grandfather paradox – the paradox of the time traveler who kills his or her grandfather in order to prevent themselves from being born.

Grand unification energy – the energy level at which all the forces are unified together.

Granulation – the appearance of the Solar photosphere due to rising convection currents from its interior.

Graticule – the grid of intersecting lines, especially of latitude and longitude on which a map is drawn or parallel lines installed in the focal plane of a telescope in order to sort out the image details.

Gravitation – the law of the universal attraction between all bodies and particles of matter in the Universe.

Gravitational field – the space surrounding a body subject to the body's gravitational influence.

Gravitational forces – fundamental forces of nature that act on bodies possessing mass.

Gravitational lensing – the bending of waves of light or radiation by a massive body lying in the line of sight between a source and the observer.

Gravitational redshift – the increase in wavelength of the light emitted by a star as it loses energy while trying to overcome a gravitational field.

Gravitational waves – waves produced by an extreme event happening in a gravitational field.

Gravitino – a subatomic particle, related to the graviton, predicted by supersymmetry.

Graviton – a hypothetical particle thought to be responsible for transferring the force of gravity.

Gravity – also called gravitation, a force that exists among all material objects in the Universe. For any two objects or particles having nonzero mass, the force of gravity tends to attract them towards each other.

Gravity tractor – a spacecraft used to deflect a celestial object from its path of collision with the Earth.

Gravity gauge duality – a concept that a Universe with gravity is equivalent to a lower dimensional Universe without gravity.

Gravity waves – wrinkles in space–time due to the change in configuration of a massive object.

Great attractor – a supposed enormous mass of matter, evolving at a considerable distance from our galaxy.

Great circle – a circle whose plane passes through the center of its own sphere.

Great dying – the most severe mass extinction in Earth's history about 252 million years ago, and probably the closest life has come to being completely extinguished.

Great red spot – an anticyclonic storm on going in the southern hemisphere of Jupiter.

Great silence – the lack of human communication with other intelligent civilizations in the Universe.

Great year – one complete cycle of the precession of the equinoxes; about 25,800 years.

Green flash – a very rare phenomenon when the Sun's uppermost structure flashes a green light.

Green goo – the idea that some form of life would eventually grow up and take over the other structures in the biosphere.

Greenhouse effect – the cause of global warming through which greenhouse gases absorb escaping terrestrial radiation and trap it close to the Earth's surface.

Greenwich Mean Time – mean solar time at the meridian through Greenwich, England.

Gregorian telescope – a type of reflecting telescope.

GRO – the Compton gamma ray observatory, the first gamma ray observatory.

Grok – the act of internalizing something by intuitive perception.

Ground state – the condition of a quantum system depleted of all its energy (at zero temperature).

GUT – Grand Unified Theory.

H II region – a zone with ionized hydrogen in interstellar space.

HA (ω) – the hour angle is one of the coordinates used in the equatorial coordinate system to give the direction of a point on the celestial sphere.

Habitable planet – a planet bearing the necessary conditions for life to exist.

Hadean – the initial period of existence of the Earth (about 3.8 billions of years) until a crust was formed, allowing conditions favourable to life to take place.

Hadron – any particle that is made from quarks, anti–quarks and gluons.

Hadron era – the time in the cosmos when the density of the Universe was dominated by hadrons.

Hagedorn temperature – the temperature in theoretical physics where hadronic matter (i.e. ordinary matter) is no longer stable, and must either "evaporate" or convert into quark matter; as such, it can be thought of as the "boiling point" of hadronic matter.

Half–life – the period of time required for half of a mass of radioactive material to decay.

Halo – a hazy shape surrounding a celestial body.

Hammer projection – a graphical representation of a sphere surface as an elliptic.

Harmonic – an overtone with a simple frequency ratio to the fundamental.

Harvard classification – a classification system for stars conceived by E.C. Pickering in the 19th century based on the absorption lines (i.e. bark lines) in the stellar spectrum, that are sensitive to the stellar temperature, rather than to gravity or luminosity.

Harvest moon – the apparition of the moon ahead of time of the autumnal equinox.

Hawking effect – the thermal radiation emitted by a black hole.

Hawking temperature – the temperature of a black hole.

Hayashi track – a luminosity–temperature relationship obeyed by infant stars in the pre-main-sequence phase (PMS phase) of stellar evolution.

Hayflick limit – the number of times a normal human cell population will divide before cell division stops.

Heat – the energy involved with a motion in the particles of a gas, liquid, or solid.

Heavy bombardment era – the period during the formation of the solar system when planetary objects were subject to continuous collisions.

Heisenberg uncertainty principle – a principle which limits the capacity of an observer to measure the position and velocity of a particle simultaneously.

Heliacal rising – the rising of a celestial object at the same time as the Sun, or its first visible rising after a period of invisibility due to conjunction with the Sun. The last setting before such a period is the heliacal setting.

Heliocentric – Sun-centred.

Heliocentric annual parallax – a method for taking the parallax of a celestial body at 6 month intervals using the diameter of the Sun's orbit as a reference mark.

Heliopause – the external limits of the solar wind.

Helioseismology – the study of the Solar interior by observations of waves at its surface.

Helium – the second most abundant element in the Universe and contains 2 protons. The most common isotope is helium 4 with 2 neutrons.

Herbig–Haro object – small patches of nebulosity associated with newly-born stars, formed when gas ejected by young stars collides with clouds of gas and dust nearby at speeds of several hundred kilometres per second.

Hertz – a unit of frequency equal to number of cycles per second.

Hertzsprung–Russell diagram – a scatter plot of stars showing the relationship between the stars' absolute magnitudes or luminosities versus their stellar classifications or effective temperatures.

Heterotic–M theory – a theory which lead to the invention of the ekpyrotic and cycle models.

Heterotic–string theory – a theory which describes all of the forces and particles of nature.

Hidden variables – an idea among some classical physicists that Quantum Mechanics isn't really as random and unmeasurable as it appears but could be entirely predictable if we knew the internal rules that were being applied. These are the hidden variables.

Hierarchy problem – a disparity of scales.

Higgs boson – an unstable hypothetical particle predicted by the Higgs mechanism, with a very short lifetime.

Higgs field – a field involved in the Higgs mechanism that prevails over the fundamental forces, and fixes the energy of the vacuum.

Higgs hierarchy problem – the problem whereby the Higgs particle has a low mass, while the particles involved with it have a high mass.

Higgs mechanism – the breaking of the electroweak symmetry.

Higgs particle – a particle carrying the Higgs field.

Higgs strength– the absolute value of the Higgs field.

Higginson – a hypothetical partner of the Higgs boson.

Hipparchus satellite – a European satellite that measures the parallax angles of 120,000 stars.

Hologram – a three-dimensional image formed by the interference of light beams from a laser or other coherent light source.

Holographic principle – a principle of string theories and a supposed property of quantum gravity that states that the description of a volume of space can be thought of as encoded on a lower-dimensional boundary to the region—preferably a light-like boundary like a gravitational horizon.

Homogeneous – the state of having characteristics that are common at all points in space.

Horizon – the line at which the Earth's surface and the sky appear to meet.

Horizon problem – the problem that the homogeneity of the Universe is not consistent with the Big Bang theory.

Horseshoe mounting – a telescope mounting for a large equatorial telescope.

Hot dark matter – hypothetical particles of dark matter which are too light and move too fast to stick together.

Hot Jupiter – a class of giant exoplanets with sizes similar to are larger than Jupiter but have very short orbital periods of less than about 10 days.

Hot Neptune – a type of giant planet with a mass similar to that of Uranus or Neptune orbiting close to its star, normally within less than 1 AU (150 million kilometers).

Hour angle – one of the coordinates used in the equatorial coordinate system to give the direction of a point on the celestial sphere.

Hour circle – a celestial circle passing through both celestial poles.

HST – Hubble Space Telescope.

HST key project – a project to measure distances to 18 galaxies through Cepheid variable stars.

Hubble classification – galaxies classified according to their shape and structure.

Hubble distance – the ratio of the speed of light divided by Hubble's constant.

Hubble's constant – the correlation between the apparent speeds of galaxies and their distances describing the current expansion rate of the Universe.

Hubble flow – the homogenous cosmic expansion.

Hubble radius – the distance at which the expansion speed equals the speed of light.

Hubble time – an evaluation of time, since the expansion of the Universe began.

Hubble ultra deep field – the deepest image taken by the Hubble telescope.

Hubble's law – the apparent recessional velocity of a celestial body is directly proportional to its distance.

Huygenian eyepiece – a lens used in small refractors.

Hydrogen – an element with one proton and one electron and is the most abundant element in the Universe.

Hydrogen 21cm emission – radio emission of 21cm wavelength generated by a transition produced by atomic hydrogen.

Hydrogen spectrum – the spectrum of atomic hydrogen.

Hydrogen–alpha line – an emission line of hydrogen in the red portion of the spectrum.

Hydrology – the study of the water cycle.

Hyper nova – a cosmic explosion, causing an immense cloudburst and emitting deadly radiation.

Hyperbola – an open curve.

Hyperboloid – the graphical representation of a hyperbola rotating about its axis.

Hypercube – a four dimensional cube.

Hyper–dimensional – involving more than the usual four dimensions of space-time.

Hyper–pyramid – a four-dimensional pyramid.

Hypersphere – a four-dimensional sphere.

Hypothesis – a scientific presentation relative to a certain phenomenon but lacking the certainty of an established theory.

I

IAU – International Astronomical Union.

IC – Index Catalogue.

ICRS – International Celestial Reference System.

Igneous rock – a cooled molten substance.

Immersion – the ingress of a celestial body into the sphere of another.

Immortality – the state of living for an infinite length of time.

Impact energy – the situation whereby, in the case of a collision between two celestial bodies, the kinetic energy of the impactor passes to the target.

Impact feature – a consequence to a celestial body of a collision with another body.

Impact rate – the frequency of impacts upon a celestial body from other celestial bodies.

Impact record – the registered computation of the impacts on a planet's surface.

Inclination – the angle between the orbital plane of a body orbiting the Sun and the plane of the ecliptic.

Inner–planet – any planet which orbits inside the main asteroid belt.

Inflation – a period of sudden rapid expansion during the early Universe.

Inflation field – the quantum field associated with inflation.

INFN – Istituto Nazionale di Fisica Nucleare, (Italian Institute for Nuclear Physics).

Infrared astronomy – the study of electromagnetic radiation falling between the red end of the visible spectrum and radio wavelengths.

Infrared light – electromagnetic radiation of a slightly longer wavelength than visible light.

Initial conditions:

(1) Physics: the state of a system at the time at which a given interaction begins.

(2) Cosmology: a quantity inserted as a given in cosmogonic equations describing the early Universe.

Instability strip – the narrow region of the Hertzsprung–Russell diagram where pulsating stars are located.

Intelligence – a trait defined by SETI as the ability and willingness to transmit electromagnetic signals across interstellar space.

Interaction – an event involving an exchange between two or more particles.

Interference pattern – a wave pattern that emerges from the overlapping and intermingling of waves emitted from different locations.

Interferometer – a device for observing the interference of waves of light or similar emanations caused by a shift in the phase of the wavelength of some of the waves.

Intergalactic matter – gas and dust in the space between galaxies.

Intergalactic medium – the hot gas that exists in the space between galaxies.

International atomic time – the international reference scale of atomic time, established by the *Bureau International de l'Heure of Paris*.

Interstellar matter – the matter contained in the region between objects in the galaxy.

Interstellar medium – the gas and dust particles between stars.

Interstellar molecules – molecules that exist in interstellar molecular clouds.

Interstellar symmetry – principle that the laws of physics do not change even when internal properties of particles do.

Interstellar travel – travel between the stars.

Intra–cluster medium – the hot gas that fills clusters of galaxies.

Invariance principle – the principle by which interactions between particles leaves them unchanged.

Inverse square law – in Newtonian mechanics, the rule that the measured intensity of light diminishes by the square of the distance of its source.

Invisible astronomy – the study of celestial objects by observing their radiation at wavelengths other than visible light.

Ion tail – the tail of a comet carried away by the solar wind.

Ionization – the process by which an atom or molecule is heated until it becomes ionized (loses its electrons).

Ionosphere – a region in the atmosphere of a planetary body in which there are free electrons and ions produced by ultraviolet radiation and x–rays from the Sun.

Irradiation – the exposure of an object to radiation.

Irregular galaxy – a non-symmetrical galaxy.

Irregular moon – a moon of a planet which is believed to have been captured by the planet's gravity and whose orbit is often large, eccentric, and highly inclined.

Irregular variable – a pulsating variable star whose variation in brightness does not follow any regular or predictable pattern.

Island Universe hypothesis – assertion that the Sun belongs to a galaxy and that the spiral nebulae are other galaxies of stars, which in turn are separated by vast voids of space.

Isotopes – atoms having the same number of protons in their nuclei but a different number of neutrons, therefore their mass differs though they may have the same number of electrons.

Isotropy – the state of having characteristics that are common in all directions.

J

Jet – a very strong outflow of energetic particles ejected by an active celestial object.

Joule – SI unit of energy.

Julian day – the continuous count of days since the beginning of the Julian Period used primarily by astronomers, and in software for easily calculating elapsed days between two events.

Jump – a transition from one state to another.

Juno – a spacecraft with the mission to survey Jupiter which entered orbit in 2016.

Jupiter – A giant planet, the largest in the solar system, and the fifth major planet from the Sun.

JWST – James Webb Space Telescope (the successor to the Hubble Telescope).

K

Kaluga – Klein theory – an array of mixed theories comprising extra dimensions involving quantum machines.

Kaons – category of particles consisting of three bosons proving that the law of physics may have small asymmetries.

Kellner eyepiece – a basic telescope eyepiece.

Kelvin – the temperature scale used throughout astronomy with absolute zero at zero Kelvin.

Kelvin–Helmholtz mechanism – a device for heating a gas which is contracting under its own gravity.

Kepler's laws – the basic laws dealing with the dynamics of celestial bodies orbiting the Sun.

Kerr's formula – the mathematical representation of a spinning black hole.

KeV – one thousand electron–volts.

Killer asteroids – asteroids which, due to their size, present a danger of a damaging collision.

Kilo – the prefix designating a factor of one thousand.

Kinetic energy – the energy a body has, during and because of its motion.

Kirkwood gaps – a particular zone in the main asteroid belt where the Jupiter resonances prevail.

Klein–Gordon equation – one of the basic equations of the relativistic quantum field theory.

Km/s – kilometres per second.

K–star – a class of cool stars with main sequence effective temperatures from 3700-5200 Kelvin and exhibiting absorption lines from ionized and neutral metals.

KT extinction – a sudden mass extinction of some three-quarters of the plant and animal species on Earth, approximately 66 million years ago.

Kuiper belt – a distinct region from the orbit of Neptune to about 1000 AU from the Sun which contains many small rocky/icy bodies.

L–star – usually a type of brown dwarf star, but very young gas giants that are still cooling from the accretion process may also have an L–type spectrum. Moreover, some hydrogen–fusing stars also fall into the L–type.

Lagrangian point – the points near two large bodies in orbit where a smaller object will maintain its position relative to the large orbiting bodies.

Landscape – the range of possible shapes the unseen dimensions of space could have.

Large Hadron Collider – the world's largest high energy particle accelerator, located at CERN, Geneva.

Large Magellanic Cloud – a satellite galaxy of the Milky Way. At a distance of about 50 kiloparsecs, the LMC is the second- or third-closest galaxy to the Milky Way, after the Sagittarius Dwarf Spheroidal and the possible dwarf irregular galaxy known as the Canis Major Overdensity.

Large scale structure – the 3-dimensional pattern of a cluster structure of galaxies.

Laser–ranger – laser used to measure distances accurately.

Last quarter – similar to the First quarter Moon phase but with the opposing half of the Moon's surface illuminated.

Latent heat – the heat that is released when a substance changes its physical state.

Late–type star – low mass stars belonging to any of the K, M, L, T, Y spectral classes.

Latitude – an angle which ranges from 0° at the Equator to 90° (North or South) at the poles.

Law – a well-established theory.

Leap second – the periodical addition of one second to the Coordinated Universal Time (UTC) in order to keep clocks throughout the world synchronized with the Earth's ever–slowing rotation.

Left-handed – biased towards turning anticlockwise.

Lens – the main piece of an astronomical instrument which turns an incoming ray of light into an image.

Lenticular galaxy – a form of galaxy in between spiral and elliptical.

Leonids – a prolific meteor shower associated with the comet Tempel–Tuttle. The Leonids get their name from the location of their radiant in the constellation Leo: the meteors appear to radiate from that point in the sky.

Lepton era – the period during which leptons dominated the cosmic density.

Lepton number – quantum number determined by the remaining quantity of leptons after the total number of anti leptons is deducted from the total number.

Leptons – subatomics particles with characteristically low masses that do not interact with the strong nuclear force.

LHC – Large Hadron Collider (Geneva).

Libration – the wagging of the Moon perceived by Earth–bound observers caused by changes in their perspective.

Lifetime – the average time ~~of~~ an unstable particle exists before it decays.

Light – electromagnetic radiation that can be seen with the naked eye.

Light cone – a point from which light paths can be seen.

Light curve – a graphical representation of the variation in light brightness of an emitting celestial object.

Limbs – the regions close to the edges of a resolved celestial object such as the Sun.

Limiting magnitude – the faintest apparent magnitude of a celestial body that is detectable or detected by a given instrument.

Lithium – a chemical element with symbol Li and atomic number 3.

Local group – the group of approximately 50 galaxies that are gravitationally interacting with the Milky Way and Andromeda.

Local symmetry – a process that can have different results in different places.

Local time – the time as determined at point on the Earth's surface in accordance with a mean solar day.

Locality – the necessity for cause and effect to occur in the same place, due to the travel time of one of the elements.

Long period comet – a comet orbiting the Sun for more than 200 years.

Long period variable – pulsating red giants with periods from 100 to 1000 days.

Longitude – points parallel to the meridian, east or west of Greenwich (England).

Lookback time – the ability to study the changes in galaxies with time by observing them at various distances, and thus different epochs making the subfield of galaxy evolution possible.

Loop quantum gravity – a version of string theory for a quantum theory of gravity.

Lorentz contraction – the apparent decrease in the length of an object along its axis of motion as seen by an observer whose speed relative to the object is relativistic.

Loss of equilibrium – a process by which the Sun increases its own temperature.

Low surface brightness galaxy – a class of galaxies that exhibit a low total luminosity and are highly extended.

LSP – Light Supersymmetric Particles.

Luminosity – a measure of the amount of energy emitted by an object per unit time.

Luminosity function – a way of determining the spread of stars as a function of their luminosity.

Lunar eclipse – the total obstruction of the moon's light by the Earth's shadow.

Lunar transient phenomenon – a temporary change in the appearance of the moon's surface.

Lunation – the mean time between two similar phases of the moon.

Lunisolar precession – the main component of precession resulting from the gravitational effect of the moon and the Sun on the Earth.

Lyman series – a series of lines in the ultraviolet part of the hydrogen spectrum, analogous to the Balmer series in the visible part.

Lyot filter – a filter to observe the Sun at a particular wavelength.

Lyrids – a meteor shower that occurs in April.

M – notation of objects in the Messier catalogue.

Mach's principle – principle that the inertia of objects comes from the distribution of mass and energy spread through the Universe.

Macho – Massive Compact Halo Object: a hypothetical object postulated to be make up dark matter.

Macksutov telescope – an improvement to the Schmidt camera which cancels out chromatic and spherical aberrations.

Macroscopic – large objects which can be seen with the naked eye.

Mandau plot – a graphical plot displaying the beginning of the Universe with stars and other bodies.

Magellanic clouds – two dwarf galaxies, close to the Milky way.

Magic number – special number of protons or neutrons within atomic nuclei that are normally stable.

Magma – molten material from within the Earth.

Magnetic field – the force field that describes the magnetic influence of electric currents and magnetized objects such as magnets or the Earth.

Magnetic moment of a particle – the ability of a particle to join in alignment with an external magnetic field

Magnetic monopole – a hypothetical particle with only one magnetic pole.

Magnetodynamics – the study of the dynamical behaviour of a plasma fluid.

Magnetosphere – the magnetic field surrounding a celestial body.

Magnification – the process of enlarging the image of an object, by an optical system.

Magnitude – the brightness of a star.

Main sequence – the region of the Hertzsprung-Russell diagram containing stars that are undergoing hydrogen fusion within their cores.

Main sequence star – a star located on the main sequence with the Hertzsprung-Russell diagram.

Mantle – the intermediary zone of the Earth's interior.

Many body problem – the complications arising when calculating the interaction of more than two objects.

Many worlds interpretation – a possible quantum mechanical solution to the problem of multiple outcomes.

Mascon – a region of the moon's surface with high density.

Massless black hole – a black hole, according to string theory, which begins with a large mass, becoming less massive as a portion of the Calabi–Yau region when it shrinks.

Mass – a physical quantity of a body that measures its resistance to acceleration under an applied force.

Mass extinction – a sharp drop in the number of species in a short period of time.

Mass–energy – the energy contained in the rest mass of an object.

Maser – a naturally occurring source of stimulated spectral line emission, typically in the microwave portion of the electromagnetic spectrum. This emission may arise in molecular clouds, comets, planetary atmospheres, stellar atmospheres, or various other conditions in interstellar space.

Materialism – the concept that material objects are the total reality of events.

Matrices – an array of numbers with their own algebraic rules.

Matrix mechanics – a version of quantum mechanics.

Matter – any substance that has mass and takes up space by having volume.

Matter era – an earlier period of time, when matter dominated the Universe.

Matter particles – the constituents of matter.

Matter waves – matter which acts as waves.

Maunder minimum – a time when only a few Sunspots are observed.

Maxwell equations – a unified theory of electromagnetism which preceeded and was supplanted by the theory of quantum electrodynamics.

Mean solar time – time correlated to the motion of the Sun traveling at a constant speed along the celestial equator.

Mean Sun – a hypothetical Sun projected as moving along the celestial equator according to the mean speed of the real Sun along the ecliptic.

Mechanics – in physics, the study of forces.

Mega – prefix for 10^6, or a million.

Megaparsec – one million parsecs.

Megaton – a unit of nuclear energy. Equivalent to 1 million tons of TNT.

Merging – the result of a collision of two galaxies.

Meridian – an imaginary circle on the surface of the Earth passing through the north and south poles.

Merlin – a multi–element radio interferometer network.

Mesons – subatomic particles that are composed of a quark and an anti-quark.

Mesosphere – the third layer in the Earth's atmosphere directly above the stratosphere.

Messenger particle – a minuscule form of a field of force.

Messier object – a celestial body listed in Charles Messier's catalogue.

Metabolic rate – quantified in time units: food consumption, heat as a form of energy, or quantity of oxygen consumed.

Metallicity – a quantification of the amount of metals (elements heavier than helium) in a star.

Metals (heavier elements) – all metals heavier than helium.

Metamorphic rock – rocks formed by pressure under high temperature from igneous or sedimentary material.

Meteor – a bright trail or streak of light that appears in the night sky when a meteoroid enters the Earth's atmosphere.

Meteor shower – a gathering of meteors in the same portion of the sky at the same time.

Meteor stream – a bulk of meteoroids orbiting the Sun.

Meteorite – a remnant of a meteoroid that went through the Earth's atmosphere from space.

Meteoroid – a relatively small body engaged in an Earth crossing orbit.

Meter – SI unit of length (about one yard).

Metonic circle – a period of 19 years in which there are 235 lunations, or synodic months, after which the Moon's phases recur on the same days of the solar year, or year of the seasons.

MeV – one mega electronvolt (1,000,000 eV).

Micro – prefix for one millionth.

Microlensing – gravitational lensing applied to smaller–sized planetary masses.

Micrometeorite – a micrometer-sized meteorite.

Micrometry – the procedure of measuring objects and the distances between them by introducing crosshairs and knife blades in the eyepiece of the sizing apparatus.

Microwave – weak electromagnetic radiation with a value of wavelength of 1mm to about 1 meter and frequencies from 1 gigahertz to 1 terahertz.

Microwave background radiation – radiation remaining from the early Universe.

Microwave light – light that has shorter wavelength than radio but longer than the far infrared.

Midnight Sun – the Sun visible at night inside the Arctic and Antarctic circles.

Milisecond pulsar – a pulsar with a magnetic field which flashes every few thousands of a second and does not decelerate.

Milkomeda (also Milkdromeda) – the projected state of the Milky Way in the time after its merging with Andromeda in a few billion years.

Milky Way – the spiral galaxy hosting the Solar System along with approximately 10^{11} stars.

Milli – prefix for 10^{-3}, or one thousandth.

Millimeter wave – the appearance of electromagnetic waves from 1 to 10 mm, at the short end of the radio spectrum.

Million – 10^6, a thousand thousand.

Minor planet – an asteroid.

Mirror – a piece of an astronomical instrument which directs the incoming light rays to the main lens.

Missing mass – invisible mass existing within galaxies and pervading clusters of galaxies.

MIT – Massachusetts Institute of Technology.

Model – a figurative conception of the actual reality, not reaching the level of a theory, just presenting facts without drawing conclusions.

Modified Newtonian dynamics – a new conception of gravity in which the gravitational forces work slower at large distances.

Moire pattern – a winding pattern end product of a set of patterns being laid one on top of another.

Molecular cloud – a cosmic cloud containing gas in the form of molecules.

Molecule – the smallest particle in a chemical compound that exhibits the chemical properties of that compound. Molecules are made up of atoms linked together by chemical bonds.

Molten – a melted solid.

Momentum – the product of an objects mass and its velocity.

MOND – Modified Newtonian Dynamics, a theory that proposes a modification of Newton's laws to account for observed properties of galaxies.

Monopole problem – the inexplicable absence of the magnetic monopole particles presumed to be from the GUT era.

Month – the necessary time for the moon to revolve once around the Earth.

Morgan–Keenan classification – stellar spectra classified according to their absorption characteristics.

Mounting – a piece on the base of a telescope allowing the apparatus to revolve.

Moving cluster – a gathering of stars spread in space that moves synchronously.

MRFM – Magnetic Resonance Force Microscopy.

M–star – a class of low mass stars with effective temperatures ranging from 2400-3700 Kelvin and exhibiting strong molecular absorption features.

M–theory – a theory formulated in 11-dimensional space–time as an extension of the superstring theory.

Multiple star – a group of stars which interact together according to their gravity.

Multiverse – the conceptual grouping of multiple Universes.

Muon – a heavier counterpart of the electron with negative electrical charge.

Muon neutrino – a heavier counterpart of the neutrino.

Nadir – the point of the celestial sphere vertically below the observer.

Nagler eyepiece – a telescope eyepiece with a large field of view.

Nano – prefix for 10^{-9}, or a billionth.

N.A.S.A. – National Aeronautic and Space Administration.

Nasmyth focus – the focus of a special telescope placed on the side of the altitude axis.

Natural – the situation wherein there are no contradictions between the theory and the real facts.

Near Earth asteroid – a minor asteroid whose orbit takes it within the vicinity of the Earth.

Nebula – an ensemble of gas and dust.

Nebular hypothesis – an early hypothesis that stars and planets are condensed from whirlpools of gas.

Negative energy – the amount of energy less than what is necessary to hold a body together.

Negative pressure – tension.

Neutralino – a hypothetical particle that appears in supersymmetric extensions of the Standard Model.

Neutrino – subatomic lepton particles with very low masses and no charge.

Neutrino astronomy – the observational study of neutrinos from astrophysical origins.

Neutron – a small particle without electric charge which is part of the atomic nucleus.

Neutron star – a degenerate stellar remnant forming from the core of massive stars and supported by neutron degeneracy pressure.

New moon – the lunar phase when the dark side of the moon faces the Earth.

Newton's constant – the numerical constant G (6.67×10^{-11} Nm^2/kg^2) in Newton's law of gravitational forces.

Newton's law of acceleration – the acceleration of a body is directly proportional to the force acting upon it.

Newton's law of action and reaction – for any action there is reaction that is equal in magnitude and opposite in direction.

Newton's law of inertia – a body continues in its state of rest, or uniform motion in a straight line, until acted upon by an outside force.

Newtonian telescope – an elementary telescope, built by Newton in 1668.

New technology telescope – a new telescope in Chile, at the European southern observatory, equipped with a system which keeps the main mirror in shape, correcting atmospheric variations.

NGC – New General Catalogue of nebulae and clusters of stars.

NMR – Nuclear Magnetic Resonance spectroscopy (a tool for chemists which helps them to study the chemical aspect of matter).

Noctilucent cloud – a cloud visible in summer at twilight hours, in the Earth's atmosphere at a high altitude.

No–hair theorem – the principle that a black hole is utterly featureless and characterized only by its mass, charge, and spin.

No–quantum Xerox principle – a theorem of quantum mechanics that forbids the possibility of a machine to copy quantum information perfectly.

Node – points at which two orbital trajectories intersect.

Non–gravitational force – any force acting on a celestial body which is not gravitational.

Non–commutative geometry – property of space–time where a zero sum game applies to the coordinates of each point. The more precise one coordinate is, the less precise the others can be.

Non–locality – the concept that objects at widely different locations are still somehow connected, even if nothing tangible stretches between them.

Non–perturbative – feature of a theory whose validity is not dependent on approximate, perturbative calculations; an exact figure of a theory.

Northern Cross – a popular name given sometimes to the cross–shaped figure of the brightness stars in the Cygnus constellation.

Northern lights – celestial lights, appearing in the skies of the northern hemisphere, and generally called the "aurora borealis".

Nova – an existing star, usually quite faint, whose brightness suddenly increases by 10 magnitudes or more, before slowly returning to its original state.

Nucleon – collective name for protons or neutrons.

Nucleosynthesis (nucleogenesis) –

(1) the fusion of nucleons to create the nuclei of new atoms.

(2) the production of chemical elements from other chemical elements via occurring natural reactions.

Nucleus –

(1) the solid part of a comet.

(2) the central region of a galaxy.

Nucleus of an atom – the centre of an atom, made of protons and neutrons.

Nutation – the periodic oscillation in the precessional motion in the Earth's axis of rotation.

O–star – a class of massive stars with effective temperatures greater than 30,000 Kelvin and exhibiting strong ionized helium lines.

OB association – a zone in the cosmos prone to produce high–mass O and B stars.

Objective – the prime lens receiving part of the light of an astronomical apparatus.

Objective prism – a prism filtering the incoming beams of light and directing them into the main lens to form the image.

Oblate spheroid – a sphere flattened at the poles.

Oblateness – the degree of deformation of a sphere.

Obliquity of the ecliptic – the angle between the plane of the ecliptic and the celestial equator.

Observable – any dynamical variable that can be seen and measured.

Observable Universe – the subset of the Universe that can be observed over some region of the electromagnetic spectrum. Note that we cannot observe the very early Universe when it was less than 400,000 years old.

Observational cosmology – the interpretation of the observed data and study of the Universe as a whole.

Observatory – an edifice built especially for celestial survey and measurement of the objects in the Universe, as well as usually hosting telescopes.

Observer – a person who is observing and recording all the activities taking place in the sky.

Occultation – temporary obstruction of a light beam coming from a celestial body, because another body passes in between the source and the observer.

Olbers' paradox – the question: *"why is the sky dark at night?"*

Omega – the parameter representing a component's density within the Universe (e.g. matter, dark matter, dark energy, etc.).

Omega factor – the result of the division of the actual density of an object by its critical density times 3.

Opera – an experiment with neutrinos at the CERN and GranSasso laboratory facilities in order to test Einstein's theory of the speed of light.

One–loop process – element involved in a computation of a couple of strings.

Oort cloud – an extended spherical shell of icy objects that exists in the outermost reaches of the solar system.

Open cluster – a group of young stars in a spiral arm of the Milky Way.

Open string – a string with both ends free.

Open Universe – a standard model which is not strong enough to stop the expansion of the Universe.

Opposition – the point in an orbit of another planet when it is opposite to the Sun on the celestial sphere.

Optical double – two stars that look very close, but not so close as to form a binary system.

Optical interferometer – a device for the study of a celestial body by interferometry at optical wavelengths.

Optics – the study of light.

Orbit – the path of a celestial body around another body.

Orbital element – a parameter used to describe an orbital such as its position or velocity at a particular time.

Order of magnitude – the exponential change in the value of a quantity.

Ordinary matter – any matter that is known to be made up of baryons.

Orionids – the most prolific meteor shower (in October) associated with Halley's Comet.

Orrery – a mechanical representation of the solar system.

Orthoscopic eyepiece – a regular telescope lens.

Oscillating Universe – an alternative to the big bang theory, in which the big bang is just part of many consecutive phases of expansion and contraction.

Oscillation – the action of some particles changing from one form to another and back again.

Oscillator – any object that goes through cyclic oscillation.

Osculating orbit – the ideal path of an orbit, if it is not disturbed by outside perturbation.

Outer planets – the planets which are orbiting beyond the main asteroid belt.

Oxymoron – two terms having opposite meaning incorporated in the same phrase, effectively countering each other.

P

PA – position angle

Pandemic – an epidemic which spreads through a region uncontrollably.

Pangaea – the name of a supercontinent that existed on Earth about 200 million years ago.

Panspermia – the transmission of life from another planet or cosmic object to the Earth.

Panstellar – related to other stars.

Parabola – an open curve.

Paraboloid – the graphical representation of a parabola being rotated about its axis.

Paradox – an oxymoron (self–contradictory proposition).

Parallax – the apparent change in the position of an object as a function of the place where it is being observed.

Parhelion – a round patch of light at the same altitude as the Sun.

Parsec – the distance at which one astronomical unit subtends an angle of one arcsecond, which corresponds to 206,265 astronomical units.

Particle accelerator — an instrument used to accelerate subatomic particles to near relativistic speeds using electromagnetic fields.

Particle decay – the normal process of matter decay.

Particle desert – the immense interval in time from the GUT transmission to the quark confinement.

Particle physics – the branch of science that deals with the smallest known structures of matter and energy.

Particles – basic units of matter or energy.

Pathogen – a biological factor - bacterial or viral - that causes illness to its host.

Patrol – a systematic survey aimed at recording certain transient astronomical phenomena.

P–brane – an object with p–space dimensions spread in space.

Penumbra – the lighter zone surrounding the shadow cast by an illuminated celestial object.

Perfect cosmological principle – the addition of an element of time to increase the value of the cosmological principle.

Peri – a prefix referring to the point in the orbit of an object at which it comes closest to its primary.

Periastron – the point nearest to a star in the path of a body orbiting that star.

Perigee – the nearest point of approach to the Earth by the moon or an artificial satellite.

Perihelion – the point in the orbit of a planet or comet at which it is nearest to the Sun.

Period – for a wave, the necessary time to complete a full cycle.

Period – the necessary time for a celestial body to complete its orbit.

Periodic comet – a common term for a short–period comet whose period is less than 200 hundred years.

Periodic table of elements – a summary of all known elements, listed by their increasing number of protons.

Period–luminosity law – a relationship between the period of light variation and the mean absolute magnitude of a Cepheid variable star.

Personal equation – a correction factor applied to all visual observations.

Perturbation theory – the application of an approximate solution to a complex problem, which subsequently will be refined.

Perturbation – the deviation of system from some state by an external force.

Phantom energy – an assumed form of dark energy.

Phase – the range of the illuminated hemisphere of a body in the solar system, as seen from the Earth.

Phase transition – evolution of a physical system between phases.

Phenotype – the compendium of properties of an organism.

Philogeny – the process of organisms evolving in time.

Photino – a hypothetical elementary particle (postulated to be a constituent of the dark matter of the universe) that theories of supersymmetry require to be associated with the photon, to have mass, and to interact only very weakly with ordinary matter.

Photoelectric effect – phenomenon in which electrons are ejected from a metallic surface when the surface is illuminated.

Photoelectric magnitude – the apparent magnitude of an object as measured with a traditional photographic emulsion which is more sensitive to blue light than the human eye.

Photoelectric photometer – an instrument that measures the brightness of a celestial object.

Photographic zenith tube – a telescope directed at the local zenith which photographs stars as they pass by.

Photon – the smallest discrete amount or quantum of electromagnetic radiation. It is the basic unit of all light and has a characteristic wavelength and energy.

Photon decoupling – the epoch of the formation of neutral atoms at which point the free electron density decreased causing a sharp decrease in opacity.

Photon–baryon gas – a gas produced in the collision of photons, protons, and electrons.

Photosphere – the visible surface of the Sun.

Photovisual magnitude – the apparent magnitude of an object measured photographically with emulsions and filters that imitate the spectral representation of the human eye.

Physics – a branch of science concerned with the behaviour of matter and energy.

Pico – multiplied by 10^{-12}.

Pinwheel Galaxy – a galaxy similar in appearance to a pinwheel (toy).

Pion – family of three bosons interacting between photons and neutrons.

Plage – a bright cloud of gas in the Sun's chromosphere

Planck length – a base unit in the system of Planck units, developed by physicist Max Planck, it is the smallest possible physical length, defined as a combination of Planck's constant and Newton's Gravitational constant. It is equal to 1.61×10^{-35} metres.

Planck time – a base unit in the system of Planck units, it is the smallest possible physical time, the time required for light to travel in a vacuum a distance of 1 Planck length. It is approximately equal to 5.39×10^{-44} s.

Planck's constant – the numerical constant h, that governs quantum phenomena

Planck's mass – the base unit of mass in the system of natural units known as Planck units. It is approximately 2.18×10^{-8} kg.

Planck's quantum principle – the concept that any wave can be emitted or absorbed in quantities proportional to their wavelength.

Planck's radiation (energy) law – the wavelength of the emitted radiation is inversely proportional to its radiation.

Planck's scale – the scale at which quantum gravity becomes important.

Planck's temperature – the base unit of temperature in a the system of Planck units, the highest possible temperature. It has a value of 1.41 x 10^{32} K.

Plane – an imaginary flat proportion of space invented for astronomical convenience.

Planet – a large and spherical non-stellar body orbiting around a star.

Planetarium – a dome which has projected onto it's interior a representation of the celestial sphere.

Planetary nebula – an emission nebula formed from the expanding outer layers around a red giant or supergiant as it evolves.

Planetesimals – bodies of solid materials formed in the earliest planetary nebulae that have grown large enough to begin to draw other materials to them by gravitational attraction.

Planisphere – a star chart analog computing instrument in the form of two adjustable disks that rotate on a common pivot.

Plasma – a state of matter without structure consisting of electrons and other charged subatomic particles.

Plate tectonics – the theory that the Earth's crust is divided into immense rocky platesand whose motions are driven by the convective motion within the mantle.

Platonic solids – one of five regular solids (a tetrahedron, cube, octahedron, dodecahedron, or icosahedron).

Platonic year – the necessary time for the celestial pole to describe a circle around the pole of the ecliptic.

Plossl eyepiece – a telescope with a wider field than an arthroscopic lens.

Plurality of worlds – a belief that there are other inhabited planets in the Universe.

Pocket Universe – a region in space that ended its rapid expansion and from this point expanded at a much slower rate.

Point – dot of zero size.

Point of no return – an analogue for the horizon of a black hole.

Polar caps – deposits of ice in the polar regions of a planet or satellite.

Polar distance – the angular distance of a celestial body from the celestial pole.

Polarimeter – an instrument measuring the polarization of electromagnetic waves.

Polarization – measure of the direction of oscillation of the electric field component of an electromagnetic wave.

Pole – either of the two points on a rotating body where its axis of rotation intersects its surface.

Pole star – a visible star, preferably a prominent one, that is approximately aligned with the Earth's axis of rotation.

Population 1 star – a relatively young star containing a high abundance of metals.

Population 2 star – an old star containing a lower quantity of metals than a population 1 star.

Population bottleneck – a special situation where adverse conditions lead a species to the brink of extinction.

Pore – a very small and short lived Sunspot.

Position angle – the orientation in the sky of one celestial body with respect to another, on a scale range from 0 to 3600.

Positron – anti–matter counterpart of the electron.

Post hoc fallacy – a fallacy in which one event is said to be the cause of a later event simply because it occurred earlier.

Post nova – the return of a nova to its original magnitude after its outburst.

Potential energy – the energy that an object has due to its mass or state in space.

Power spectrum – the distribution of the energy of a waveform among its different frequency components.

Poynting–Robertson effect – a non–gravitational force produced by the action of solar radiation on small particles in the solar system which causes them to spiral inward towards the Sun.

Pre–big bang scenario – a cosmological model based on string theory in which time extends back before the advent of the big bang.

Precession – the circular motion of the celestial pole, around the poles of the ecliptic, and the associated westward movement of the equinoxes with respect to the background stars.

Preciding – the leading edge, feature, or member of an astronomical object.

Pressure – the force exerted against any obstacle, relative to the contact area of the object involved.

Primary – the object around which another body is orbiting, or the main lens of a telescope.

Prime focus – the point at which the objective lens or primary mirror of a telescope brings the oncoming light to a focus.

Primeval fireball – the explosion which, according to the big bang theory, resulted in the present Universe.

Primordial – term used to refer to the chemical or physical condition prior to the big bang.

Primordial black hole – a black hole remnant from the early Universe.

Primordial nucleosynthesis – the production of atomic nuclei, in the first three minutes after the big bang.

Principle – a statement that describes the rules of behaviour of the natural world.

Principle of equivalence – acceleration and the force of gravity are the same.

Principle of relativity – core principle of special relativity claiming that all constant velocity observers are subject to an identical physical law, and therefore all constant velocity observers are justified when claiming to be at rest.

Prism – a device which splits incident light into its basic frequencies.

Probability amplitude – in quantum mechanics, a complex number whose absolute value squared gives a probability.

Probability interpretation – the concept that a wave function allows only the probability of finding a particle at a particular location.

Problem of frozen time – a conundrum that according to the general theory of relativity, the world should be static and unchanged.

Prokaryote – a unicellular organism, sometimes a multi–cellular organism that lacks a membrane–bound nucleus, mitochondria, or any other membrane–bound organelle.

Prominence – a cloud of matter protruding from the Sun's chromosphere into the corona.

Proper motion – the apparent motion of a star on the celestial sphere, as a result of its movement relative to the Sun.

Proper time – the time elapsed according to a moving clock; a measure of distance along a world line.

Proplyd – a syllabic abbreviation of an ionized protoplanetary disk, an externally illuminated photo–evaporating disk around a young star.

Proportional – a value which is related to another one, by the fact that no matter in what mathematical operation one of them is involved, both are affected.

Protogalaxy – a galaxy in the process of formation.

Proton – a subatomic particle, symbol p or p+, with a positive electric charge of +1e elementary charge and a mass slightly less than that of a neutron.

Proton decay – the normal disintegration of the proton, predicted by the grand unified theory.

Proton–proton reaction – the nuclear reaction that fuses hydrogen into helium and is responsible for the source of energy for main sequence stars with masses equal to or less than that of the Sun.

Protoplanet – the early stages of planet formation by the accretion of planetesimals and/or gas.

Protostar – a young star formed from dense clusters of gas collapsed into molecular cloudsbut not yet contracted onto the main sequence.

Pulsar – a rapidly rotating neutron star that is highly magnetized such that it beams radiation and appears to pulse.

Pulsating star – a kind of variable star which fluctuate in brightness.

Purkinje effect – the change in the perception of the human eye from yellow towards green in certain cases of transition from light to obscurity.

QCD – Quantum Chromo-Dynamics.

QCD strings – strings made of gluons that bind quarks together to form hadrons.

Quadrant – an astronomical tool that is used for measuring the angle of a star above the horizon.

Quadrantids – a meteor shower.

Quadrature – any celestial body positioned at 90° relative to the Sun.

Quantum chromodynamics – theory that for colour and electric charge quarks and gluons are respectively responsive.

Quantum electrodynamics – the quantum theory of electricity and magnetism.

Quantum field – a collection of particles interacting in the same space.

Quantum field theory – the quantum mechanical theory of fields.

Quantum foam – quantum mechanical fluctuations of space time at Planck scales.

Quantum fluctuation – variations of a system's properties over short time intervals.

Quantum gravity – a tentative attempt to unify quantum mechanics and general relativity.

Quantum mechanics – a probabilistic method of obtaining information about the physical world through wave functions.

Quantum number – a description for a stored quantity.

Quantum physics – the section of physics that studies the actions of atomic or subatomic scale systems.

Quark – a fermion that is under the influence of the strong nuclear force.

Quark confinement – the interval in time (10 microseconds) when quarks join protons, neutrons, mesons, and their antiparticles.

Quark era – the period in the evolution of the early universe when the fundamental interactions of gravitation, electromagnetism, the strong interaction and the weak interaction had taken their present forms, but the temperature of the universe was still too high to allow quarks to bind together to form hadrons.

Quark–gluon plasma – a phase near nuclear density in which protons and neutrons mix with quarks and gluons.

Quark number – a certain quantity resulting from deducting the number of actual quarks from the number of given antiquarks.

Quasar – a very large and powerful active galactic nucleus emitting radiation across a range of wavelengths.

Quintessence – an alternative to vacuum energy, sometimes considered a fifth force, and also known as dark energy. A dynamical field with an energy density that slowly changes.

R

RA – right ascension.

Radar astronomy – the use of radar to study objects in space.

Radial velocity – the velocity of an object coming to or from an observer.

Radiant – for a meteor shower, the celestial point in the sky from which (from the point of view of a terrestrial observer) the paths of meteors appear to originate.

Radiation – any form of energy emitted through waves.

Radiation belts – region of a planet's magnetosphere in which charged particles are trapped by the planet's magnetic field.

Radiation pressure – pressure exerted by electromagnetic radiation.

Radiative zone – the region within a star's interior in which energy is most efficiently transported via the the propagation of photons.

Radio astronomy – a branch of observational astronomy that deals with the detection of radio photons.

Radio galaxy – a galaxy emitting an unusually high number of radio waves.

Radio interferometer – a radio telescope that operates as an optical interferometer.

Radio telescope – a telescope meant to receive and record all radio waves coming from space.

Radio waves – electromagnetic waves with long wavelength.

Radioactivity – a process in which an unstable nucleus disintegrates in order to form a stronger one by acquiring a better configurationby emitting radiation.

Radiocarbon dating – determination of the age of objects containing radioactive carbon by means of its radioactive half–life.

Radiometric dating – determining the age of objects by means of the half–life of the unstable elements they contain.

Radius vector – a conceived line from an orbiting celestial object and its primary.

Ramsden eyepiece – a simple telescope eyepiece made of two basic parts.

Rare Earth – the idea that conditions conducive to intelligent life are very rare, and thus the chances of finding another world like ours is slim.

Rarefied – a term specific to vacuous gases.

Rays – a model representation of light that points in the direction of propagation of the wave.

Realism – the new philosophy that proposes the view that there is a reality independent of the observer.

Recurrent nova – a nova that erupts periodically.

Recession velocity – the speed at which celestial bodies are moving away from each other.

Recombination – the era in the early Universe at which point the temperature had become sufficiently cold to enable electron capture by proton.

Red dwarf – a class of cool and low mass main sequence stars with effective temperatures 2400-3700 Kelvin and exhibiting strong molecular absorption.

Red giant – a giant star of spectral type K or M, having a surface temperature of less than 4700k, a diameter 10 to 100 times that of the Sun, and a luminosity 100 to 10,000 that of the Sun.

Red planet – the planet Mars, which appears red in color.

Reddening – a phenomenon of increasing the intensity of red photons from a distance source due to the preferential scattering of blue photons by interstellar dust.

Redshift – a stretching of the wavelength of radiation from a source due to the fact that the source is moving away or due to the expansion of the Universe.

Redshift distance relation – the correlation between the redshift in the spectra of galaxies and their distances.

Redshift survey – a graphical representation of the scattering of galaxies in the cosmos.

Reductionism – a process by which complex phenomena are separated in small segments in order to facilitate the way of dealing with.

Reflecting telescope – a telescope in which the image is formed in the back of the tube through a secondary mirror and sent to a focus in a particular part of the instrument.

Reflection nebula – a nebula which has the dust and gas particles within reflecting the starlight.

Refracting telescope – a telescope that forms an image from the refracted light from a lens situated in front of the tube.

Regolith – the surface dust and debris covering bodies in the solar system.

Regression of the nodes – the slow westward movement of the nodes of the moon due to the gravitational pull of the Sun.

Regular moon – a moon of a planet that likely formed from the disk of circumplanetary material shortly after the formation of the planet.

Reheating – the transformation of scalar vacuum energy into radiation and particles.

Relative Sunspot number – the averaging of Sunspot numbers seen by different observers, in order to arrive at a common figure.

Relativistic – a term used to describe particles moving near to the speed of light.

Relativity theory – Einstein's theory of motion and gravity, comprising the special theory of relativity and the general theory of relativity.

Riemannian geometry – calculus dealing with curved shapes of all proportions.

Renaissance – the period of enhancement in the west of Europe starting in 1350 and ending in 1600 or with the death of Shakespeare in 1616.

Renormalizability – adjusting all scale value processes to a common denominator so the basic reaction value of a force is the same at all scales.

Renormalization – eliminating the negligible infinitesimal elements from quantum mechanics by a mathematical process.

Residuals – the discrepancy between anticipated and actual observed values.

Resolution – the degree of visibility of the details in an image.

Resolving power – the capacity of an optical system to discern two objects being close but apart.

Resonance – a product of the gravitational interaction of two bodies orbiting the same primary.

Rest energy – the inherent energy stored within a body with a non-zero mass.

Retrograde motion – orbital or rotational motion of a celestial body in the opposite direction to the Earth's movement.

Revolution – the movement of a celestial body along its orbit while it is rotating around its axis.

RHIC – Relativistic Heavy Ion Collider.

Rich–field telescope – a low powered telescope equipped with a wide–angle eyepiece.

Right ascension – a geodesic factor utilized to pinpoint the position of a celestial body on the celestial sphere.

Right handed – one of the two mirrors image of a particle related to how it responds to the weak nuclear force.

Ring galaxy – a galaxy in the form of a ring around the nucleus.

Ring (planetary) – a narrow ring or thin disk of dust particles orbiting in the equatorial plane of a planet.

RNA world – a transitional period of time, preceding DNA, when RNA organized genetic components and catalyzed its own replication.

Roche limit – the distance away from a small body at which point the tidal force from a nearby large body can pull the small body apart.

ROE – Royal Observatory, Edinburgh.

Roque de los muchachos observatory – an observatory in LaPalma, in the Canary Islands.

Rotation – the spinning of an object around its own axis.

Rotation curve – the orbital velocity of objects as a function of their distance from the primary.

Royal Greenwich Observatory – an English observatory in London.

Royal Society – an organization founded in England in the 17th century, for the advancement of science.

R–process – fast aggregation of neutrons in stellar nucleosynthesis.

Runaway star – a star of spectral type O, or B, which follows an irregular orbit in space.

RV Tauri star – a star in the constellation Taurus. It is a yellow supergiant and is the prototype of a class of pulsating variables known as RV Tauri variables.

S

Sachs–Wolfe effect – fluctuations in the cosmic microwave background due in part to a combination of gravity and cosmic expansion.

Saros – a period of approximately 223 synodic months that can be used to predict eclipses of the Sun and Moon.

Satellite – a celestial body in orbit around a planet.

Scalar field – a field that requires only one coordinate.

Scale factor – a factor concerning the size of the Universe in the process of expanding, mainly the space between the celestial bodies.

Scale invariance – the property whereby characteristics of an object are not affected when multiplied by a common factor.

Scale invariant spectrum – a spectrum which has the same height waves even under scale invariance.

Scarp – a line of cliffs formed by the faulting or fracturing of the Earth's crust.

Scattered disk objects – Kuiper belt disk objects.

Scattering – process where some forms of radiation (such as light), sound, or moving particles, are forced to deviate from a straight trajectory by one or more paths due to localized non–uniformities in the medium through which they pass.

Schmidt camera – is a catadioptric astrophotographic telescope designed to provide wide fields of view with limited aberrations.

Schmidt–Cassegrain telescope – a catadioptric telescope that combines a Cassegrain reflector's optical path with a Schmidt corrector plate to make a compact astronomical instrument that uses simple spherical surfaces.

Stochastic cooling – a form of particle beam cooling. It is used in some particle accelerators and storage rings to control the emittance of the particle beams in the machine. This process uses the electrical signals that the individual charged particles generate in a feedback loop to reduce the tendency of individual particles to move away from the other particles in the beam.

Schrodinger's cat – a hypothetical cat that is dead and alive at the same time.

Schrodinger's equation – the basic equation that deals with the behaviour of particles according to wave mechanics.

Schrotter effect – a discrepancy between the projected phase Venus should follow and the observed one.

Schwarzschild radius – a physical parameter that shows up in the Schwarzschild solution to Einstein's field equations, corresponding to the radius defining the event horizon of a Schwarzschild black hole.

Science – systematic study of nature.

Scintillation – the apparent change of the stars in brightness and colour, when observed with the naked eye, due to changes in the Earth's atmosphere.

Sea–level air pressure – the pressure exerted by a column of air from the atmosphere over the surface of the sea.

Season – a sequence of the surface conditions of a planet due to its axial revolution as it is orbiting around the Sun.

Second – the SI unit of time.

Second contact – in a total solar eclipse, the point when the moon totally covers the Sun's disk.

Second law of thermodynamics – a physical law stating that entropy always increasesin any thermodynamical process.

STEP – a Cassini mission testing general relativity.

Secondary – a celestial body which orbits around another called the primary.

Secular parallax – the movement of a body in time, a consequence of the Sun's orbit.

Secular acceleration – the slowly increasing speed of the moon orbiting the Earth.

Sedimentary rock – rock that has grown over time with deposits of organic or mineral matter.

Seeing – the degree of accuracy of an image from a telescope, considering the atmospheric conditions.

Seismology – the study of earthquakes and their consequences.

Solectron – the super symmetric partner of the electron.

Selenography – the study of the surface and physical features of the Moon.

Self–replicating space probes – probes that can travel to nearby stars, make replicas of themselves and explore the galaxy.

Semi–major axis – the length of half of the long axis of an elliptical orbit.

Semi–regular variable – a pulsating star with a period of 20 to 2000 days or more and amplitude from a few hundred to a magnitude of several thousand.

Senescence – complex biological processes related to aging.

Separation – the distance between the members of the same star system.

SETI – Search for Extra–Terrestrial Intelligence.

Sextant – instrument used in navigation and for locating places in general in the 19th century.

Shadow bands – light coloured streaks appearing in the sky before and after a total eclipse of the Sun.

Shadow matter – matter composed of hypothetical particles which are supposed to exist by virtue of the super symmetric theory.

Shell star – a hot B–type star with an equatorial disk thrown out by rapid rotation.

Shepherd moon – a smaller moon that controls through its gravitational power the minor particles orbiting in a planetary ring.

Shooting star – a meteor.

Short–period comet – a comet which appears frequently allowing observers to register its orbital characteristics.

Short–wavelength x–rays – electromagnetic radiation longer than gamma rays.

Sidereal day – the necessary time for a star to cross the observer's meridian twice.

Sidereal period – the orbital period of a planet.

Sidereal time – local time in relation to the rotation of the Earth with respect to the stars.

Sidereal year – the period of time in which the Earth completes a revolution around the Sun.

Siderostat – a mirror fixed on a telescope in order to compensate for the movement of the targeted celestial object, by directing the incoming beam of light into the main lens of the telescope.

SIM – Space Interferometry Mission.

Simulation hypothesis – the concept that we live in an alternative world created by a superior intelligence.

Simultaneity – different events that take place at the same time.

Synchotron radiation – the electromagnetic radiation emitted by charged particles which are accelerated to speeds just under the speed of light.

Synchronous rotation – the period of revolution and the axial rotation of a celestial body turning simultaneously.

Singularity – a place in space-time where the regular laws of physics as we know them do not apply.

Singularity theorem – a theory that states that, under certain conditions a singularity must take place, for example the big bang.

SIRTF – Space Infrared Telescope Facility.

SLAC – Stanford Linear Accelerator Center.

Slepton – a hypothetical boson superpartner of a lepton whose existence is implied by supersymmetry.

Sloan digital sky survey – a survey meant to observe a quarter of the sky.

Slow roll – a term used to depict the deployment of the scalar field during the inflation phase of the Universe.

Small Magellanic cloud – a prominent dwarf galaxy near the Milky Way.

S–matrix – a unitary matrix in quantum mechanics, the absolute values of the squares of whose elements are equal to probabilities of transition between different states.

Snowball Earth – an era 700 millions ago when a change in temperature caused the Earth to be covered by ice.

SNR – Supernova remnant.

SOFIA – Stratospheric Observatory for Infrared Astronomy.

SOHO – Solar and Heliospheric Observatory.

Solar constant – the value of the power received per unit area from the Sun at the Earth's distance (equal to 1367 W/m^2)

Solar cycle – the cyclic occurrence of Sunspots over a period of 11 years.

Solar day – the duration of time during which the Sun crosses the meridian twice.

Solar eclipse – an astronomical even whereby an observer passes through the shadow cast by the Moon fully or partially blocking the Sun.

Solar flare – an abrupt, enormous release of energy from the Sun.

Solar luminosity – total radiative power of the Sun.

Solar mass – the mass of the Sun used as comparison for measurements of other bodies in the solar system.

Solar maxima – the time when the Sun is at its highest level of activity, shown by its sunspots.

Solar minima – the time when the Sun is at its lowest level of activity, when it shows no spots.

Solar system – the collection of objects (planets, asteroids, Kuiper belt object, etc.) that are gravitationally bound to the Sun.

Solar wind – the steady flow of charged particles from the solar corona into space.

Solstice – the most northern or southern position of the Sun relative to the meridian.

Solstice colure – an imaginary circle going through the north and south celestial poles as well as the winter and summer solstice.

Southern–cross – the common appellation of the southern hemisphere constellation Crux.

Space – the boundless three–dimensional extent in which objects and events have relative position and direction.

Space colonization – the concept of continuing human life in space.

Space interferometry mission – an envisaged space mission intended to make accurate measurements of the position of celestial objects.

Space invader – an object coming from far in the Universe that can penetrate our space.

Space probe – an unmanned vehicle propelled in space to observe certain parts of the Universe.

Space velocity – the true velocity of a star in space with respect to the Sun.

Space–time – the unification of space and time according to Einstein.

Space–time diagram – a graph of space and time tracking the orbits of objects in space and time.

Space–time foam – erratic behavior of space–time on the smaller scales, caused by the quantum fluctuations of the gravitational field.

Sparticle – is a class of hypothetical elementary particles. Supersymmetry is one of the synergistic theories in current high-energy physics that predicts the existence of these "shadow" particles.

Spatial dimension – the three dimensions of space.

Special theory of relativity – Einstein's theory of motion in the absence of gravity.

Species – an association of individuals, living together, and able to reproduce themselves.

Spectral classification – categorizing the stars according to their spectra.

Spectral energy distribution of blackbody radiation – a blackbody spectrum.

Spectral line – a bright or dark feature in a spectrum resulting from either emission or absorption of photons.

Spectral type – a classification of stars by their spectral lines.

Spectrograph – a device that performs the function of a prism by splitting incident light into its constituent wavelengths to produce a spectrum.

Spectroscope – an instrument designed to present spectra for the analysis of electromagnetic radiation.

Spectroscopic binary – a process of analyzing the spectra of a binary, when two stars are orbiting so close together like one sole body.

Spectroscopic parallax – the position of a celestial body deduced by examining a star spectrum.

Spectrum – a band of the components of the emissions of electromagnetic radiation, classified according to their wavelength.

Speed of light – the maximum possible velocity (300,000 kms^{-1} in a vacuum).

Spherical aberration – an image defect caused by a deformed lens which averts the formation of a proper focus.

Sphere – the outer surface of an object in the form of a ball, where all points on the surface are at equal distance from the centre.

Spheroid – the surface resulting from rotating an ellipse around one of its axes.

Spicule – a constricted jet coming out of the Sun's chromosphere.

Spin – a quantum characteristic of elementary particles (rotational motion).

Spin network – a diagram connecting all special events.

Spiral galaxy – a galaxy consisting of a nucleus of stars with spiral arms emerging, winding around the nucleus and forming a flattened, disk shaped region.

Spiritualism – the concept that some phenomena are beyond our comprehension, and we have to accept them as such.

Spontaneous emission – the emission of a photon as an atom makes the transition from an excited state to a lower energy state

Sporadic meteor – a meteor which orbits independently of recognized meteor showers.

Sporer's law – principle governing the appearance of sunspots at lower latitudes during the 11 year solar cycle.

S–process – slow join of neutrons in stellar nucleosynthesis.

S–process elements – special elements made during the s process.

Sterile neutrinos – hypothetical particles that interact only via gravity and do not interact via any of the fundamental interactions of the Standard Model.

String theory – a theory which proposes the idea that all components of the Universe are made of vibrating strings uniting all particles behaving as a single set.

Squark – hypothetical super symmetric partner of the quark.

Squid – superconducting quantum interference device (used for detecting the quantum state, and for manipulating quantum information).

SS433 – an unusual binary star, number 433 in the catalogue of stars with bright emission lines.

S–star – a giant star with a surface temperature similar to that of an M giant, with a spectrum containing absorption bands of zirconium and titanium oxide characteristic to M stars.

Stable – remaining in a constant state.

Standard candles – objects of known luminosity used as distance indicators by measuring their brightness.

Standard epoch – a period of time chosen as the beginning of an astronomical era wherein the positions of all the stars are recorded in order to avoid the effects of precession and proper motion together with a system of reference in which every position is measured.

Standard model – the theories of the four forces, which, if taken together, can predict the outcome of every known fundamental interaction.

Standard model of particle physics – the modern theory of particle physics and their interaction.

Star – a celestial body that generates energy by means of nuclear fusion in its core.

Star evolution – the development of a star, during its existence.

Star cluster – a gathering of stars loosely bound by gravity and moving together through space.

Star streaming – the moving order of stars in their orbits around the galaxy.

Star trailing – in astrophotography, the images of stars and other objects appear deformed as trailing if taken by an apparatus which is not equipped to compensate for the Earth's rotation.

Starburst galaxy – a galaxy going through an intense star–birth period.

Stark effect – the splitting of spectral lines when atoms are placed in an electric field.

Stationary point – the point in the orbit of a planet were it looks motionless as seen from Earth.

Stationary state – the theory that this is a sphere spinning at the same speed and not changing in time.

Steady state theory – a theory that the Universe has always existed and has neither beginning nor end and always stays the same.

Stellar corpse – the remnants of a star after its nuclear fusion ends.

Stellar evolution – the evolving process of stars over time.

Stellar mass loss – the expulsion of matter from the peripheral layers of a star due to its own generated radiation.

Stellar nomenclatures – naming system using the names of constellations to identify the stars within them.

Stellar populations – a catalogue of stars according to their era and place in the galaxy.

Stellar wind – a flow of charged particles from the surface of a star.

STEP – Satellite Test of the Equivalence Principle.

Sterilizing impact – an impact of a celestial body onto a planet capable of wiping out all life on that planet.

Stimulated emission – when a photon is not absorbed by an excited atom, but stimulated to emit a second photon of the same frequency.

Stochasting cooling – a form of particle beam cooling. It is used in some particle accelerators and storage rings to control the emittance of the particle beams in the machine.

Stony meteorite – a rock–like meteorite made of a compound of silicates with some chemically uncombined metals.

Stony–iron meteorite – a meteorite composed of silicates, metals, and minerals, but mostly nickel and iron.

Storage ring – a device keeping particles in circular motion in a laboratory, before they are injected into the main accelerator ring.

Strange quark – a second generation massive quark with negative one third of the elementary charge.

Stratosphere – the second layer in the Earth's atmospherein which the temperature increases with height.

String – an extended one-dimensional object thought to be the basic building blocks of nature.

String coupling constant – a number that determines the point at which a string should separate into two strings or for two strings to unite to form one string.

String theory – a theory which proposes the idea that all components of the Universe are made of vibrating strings uniting all particles behaving as a single set.

Stromatolites – fossilized microbial residues from 3.5 billion years ago which have descendants thriving today.

Strong lensing – gravitational lensing that is strong enough to produce multiple images of the background lensed object.

Strong nuclear forces – a basic force of nature that keeps quarks together and holds nucleons together as nuclei of an atom.

Structure problem – the incapacity of the big bang theory to supply the necessary amount of material to later form stars and galaxies.

Subatomic – objects on a smaller scale than an atom.

Subduction – the action of a tectonic plate slipping over or under another tectonic plate.

Sub–dwarf – a star with lower luminosity than other stars of the same spectral type.

Sub–giant – a star with lower luminosity and smaller radius than stars of the same spectral type.

Sub–horizon growth – density variations smaller than the horizon size growing in amplitude because gravity pulls matter from low density regions to high density regions.

Sublimation – the process of transitioning from the solid state to the vapor state, bypassing the liquid state.

Submillimeter wave astronomy – astronomical observations using electromagnetic waves at wavelengths of less than one millimetre.

Sub–solar point – the point on the surface of the Earth which is directly under the Sun at a given moment.

Sum over histories – presumed reconstruction of past conditions, considering quantum possibilities and reconsidering all contingencies.

Summer triangle – the triangle formed by the three top magnitude stars: Altair, Vega, and Deneb. This triangle is visible on summer nights in northern temperate latitudes.

Sum–over paths – a presumed situation in quantum mechanics when particles are free to move in every way conceivable.

Sun – the main star in the solar system around which all the other celestial bodies revolve.

Sundial – a crude timekeeping device which indicates local time.

Supernova – the explosion of a massive star that has reached the end of its life.

Sun–grazer – a comet which intersects the Sun's corona at perihelion.

Sunrise – the instant when the Sun appears over the horizon.

Sunset – the instant when the Sunsets below the horizon.

Sunspot – a darker region in the Sun's photosphere, cooler than its surroundings.

Sunyaev–Zeldovich effect – the distortion of the Cosmic Microwave Background radiation through inverse Compton scattering by high energy electrons in galaxy clusters, in which the low energy Cosmic Microwave Background photons receive an average energy boost from collision with the high energy electrons of the cluster.

Super flares – instances where a massive star exhibits much stronger coronal activity than normal.

Super–Kamiokande – a neutrino observatory located under Mount Ikeno near the city of Hida, Gifu Prefecture, Japan.

Super cluster – very large clusters of galaxies.

Superconducting supercollider – an accelerator of huge size and energy.

Super–Earth – an Earth-type exoplanet much bigger than the Earth.

Supergiant – a very luminous star with a large diameter and low density.

Supergranulation – the granulation pattern observed on the surface of the Sun.

Supergravity – a tentative attempt to join super symmetry with the general theory of relativity.

Super–horizon growth – the way in which density larger than the horizon size grows in amplitude because of differential expansion.

Super lattice – an assemblage of periodic layers of divergent semiconductors producing a semiconductor laser emitting coherent pulses of radiation by a process of quantum cascade.

Superluminal – the motion of an object which apparently moves faster than the speed of light.

Supermassive black hole – a black hole much larger than normal, residing in the core of a galaxy.

Supernova – the explosion of a star in which the star may reach a maximum intrinsic luminosity one billion times that of the sun.

Supernovae remnant – a nebula of gas, ejected into space by a supernova.

Superpartner – in supersymmetry, the particle corresponding to a similar established particle with the opposite electric charge.

Superposition – in quantum theory, a multitude of possible results from a quantum state composed of two or more other states.

Superstring – a string that is super symmetric.

Superstring theory – the theory that elementary particles of matter are represented by tiny vibrating strings.

Super–symmetry – a principle that proposes a relationship between two basic classes of elementary particles: bosons, which have an integer–valued spin, and fermions, which have a half–integer spin.

Superunified theory – a theoretical assumption which can demonstrate how in the early Universe all four forces of nature acted together.

Surface tension – the tension at the surface of a liquid caused by the attraction of the liquid molecules towards one another.

Surveyor – a series of robotic lunar lander spacecraft launched in 1968.

Suspended animation – the process of slowing a metabolism to a state near to arrest, in order to have it survive a long recess.

Symbiotic variable – a binary star exhibiting a large difference in surface temperature between the two component stars.

Symmetry – the principle that properties of objects or equations do not change even if they are involved in transformations.

Symmetry breaking – the situation when the symmetry of a theory is hidden from view by the dynamic of the system.

Symmetry group – a mathematical group that unites its components and shows a symmetry.

Synchronous orbit – an orbit in which the period of revolution of a celestial object is the same as the period of its axial rotation.

Synchronous rotation – an occurrence whereby the axial rotation of a celestial orbit take place in the same period as its period of revolution.

Synodic month – the time interval between two similar phases of the moon.

Synodic period – the cyclic period of one body around another, as seen from the Earth.

Syzygy – a straight–line configuration of three or more celestial bodies in a gravitational system.

T

Tachyon – a hypothetical particle that always travels faster than light.

TAI (temps atomique international) – International Atomic Time.

Tail – a lengthy trail composed of gas particles and cosmic debris ejected from a comet nucleus.

Tau – an elementary particle similar to the electron, with negative electric charge and spin.

Tau neutrino – a subatomic elementary particle which has the symbol ντ and no net electric charge.

TDFS – Two Degree Field Survey (Australia).

T–duality – in string theory, the equivalence between a small dimension of space and a large one.

Tectonics – the deformation of a planetary exterior due to interior heating.

Tektites – minuscule glassy objects found spread over some regions of the Earth.

Teleomere – repetitive DNA at the end of chromosomes which prevents them from having cancer or destroying themselves.

Teleportation – instant transfer of information from one place to another using entangled particles.

Telescope – optical instruments that make distant objects appear magnified by using an arrangement of lenses or curved mirrors and lenses, or various devices used to observe distant objects by their emission, absorption, or reflection of electromagnetic radiation.

Telescope reflector – a telescope using mirrors to guide the incoming beam of light.

Telluric line – an absorption line in the spectrum of a celestial object produced by molecules in the Earth's atmosphere.

Temperature – a measure of the average vibrational energy stored in an ensemble of particles.

Tension – the pulling force transmitted axially by the means of a string, cable, chain, or similar one–dimensional continuous object, or by each end of a rod, truss member, or similar three–dimensional object.

Termination shock – the moment in the heliopause when the solar wind goes from subsonic to supersonic speed.

Terminator – from the Earth, the view of the separation of day and night as the planet rotates.

Tera – trillion (x 10^{12}).

Terraforming – the idea of changing the climate of a planet, in order to transform make it inhabitable for humans.

Terrestrial Planet Finder (TPF) – a NASA project designed to find Earth–like planets, which may have life on them.

Tesserae – regions of heavily deformed terrain on Venus, characterized by two or more intersecting tectonic elements, high topography, and subsequent high radar backscatter.

TeV – trillion electron volts (unit of energy).

Tevatron – a circular particle accelerator (inactive, since 2011) in the United States, at the Fermi National Accelerator Laboratory (also known as Fermilab), east of Batavia, Illinois, holding the title of the second highest energy particle collider in the world, after the Large Hadron Collider (LHC) of the European Organization for Nuclear Research (CERN) near Geneva, Switzerland.

Thanatology – the scientific study of death and the losses brought about as a result.

Theory – a well–substantiated explanation of some aspect of the natural world, based on a body of facts that have been repeatedly confirmed through observation and experiment.

Theory of everything – a theory which tries to unite all four forces of nature in one single whole.

Thermal energy – the energy contained within an object, due to its atomic or molecular vibrations.

Thermal spectrum – the spectrum of radiation produced by a hot object.

Thermodynamics – the study of heat and its relations to other forms of energy. the field of heat and its effect.

Thermonuclear – nuclear processes that liberate great amounts of heat energy.

Thermonuclear fusion – a reaction in which nuclei collide and unite releasing huge amounts of thermal energy.

Thermosphere – the fourth layer in the Earth's atmosphere which absorbs the energy radiated by the Sun as the temperature increases.

Theta–plus – a subatomic particle made of five quarks.

Third contact – the instant when, in a solar eclipse, the moon starts revealing the Sun.

Thompson scattering – a procedure by which photons are scattered by free electrons.

Thought experiment – a virtual experiment conceived in order to verify the validity of a physical theory or concept.

Three–body problem – the problem of how to determine the behaviour of three bodies moving only under their mutual gravitational attraction.

Tidal forces – the differential force of gravity acting with different strengths at various points of an object thus causing a deformation of the object.

Tides – changes made on the surface of a celestial body by the gravitational force of attraction of other bodies.

Tilt – the deflection from a simple primordial roughness spectrum.

Time – a factor used to measure the duration of events.

Time dilation – the slowing of the passage of time for a moving observer.

Time zone – a graphical division of the Earth's surface on paper in order to determine the local time.

Top quark – a massive third generation quark with two thirds of the elementary charge.

Top–down approach – a method in cosmology of studying the evolution of the Universe backwards.

Topocentric – observations made from a certain point on the Earth.

Topography – a map of elevation of a celestial body's surface.

Topology – the study of geometrical properties and spatial relations unaffected by the continuous change of shape or size of figures.

Topology–changing transition – development of the cosmic background that entails shredding or tearing apart, and consequently the set–up of the Universe.

Torus – a surface of revolution generated by revolving a circle in three-dimensional space about an axis coplanar with the circle.

TPF – terrestrial planet finder (a NASA project).

Transhumanism – a global movement advocating the use of technology in order to enhance human competency.

Transit :

(1) the passage of a celestial body across the observer's meridian.
(2) the passage of a body directly between the Earth and the Sun.
(3)the passage of a planetary satellite across a planet's disk.

(4) the passage of a surface or atmospheric feature of a body across its central meridian as it rotates.

Transit instrument – a telescope that glides only in one plane mounted on a horizontal east–west axis.

Transition region – a slim zone of 100 km between the corona and the chromosphere of the Sun's atmosphere.

Translation – in physics, the displacement of an object from one location to another without turning it.

Trans–Neptunian object – any minor planet in the Solar System that orbits the Sun at a greater average distance than Neptune, which has a semi-major axis of 30.1 astronomical units (AU).

Threshold temperature – the temperature above which energy particle collisions can create new particle–antiparticle pairs.

Triangulation – a way of determining the distance to an object.

Trigonometric parallax – the method of measuring the angular displacement of a celestial object from different vantage points as the Earth orbits the Sun in order to measure the distance to that object.

Trigonometric parallax – a method of finding the distance to an object. [repeat]

Trillion – a thousand billion (10^{12}).

Triple alpha process – a nuclear reaction which converts three alpha particles (helium nuclei) into carbon with a release of energy as a result.

Triplet – a combination of three lenses which are joined closely with a very small space in between.

Trojan asteroids – small rocky objects whose orbits lead or trail a planet in the Solar System at the same distance from the Sun.

Tropic of Cancer – the parallel of 23.5 degrees of latitude north of the Earth's equator that determines the extreme possible northern declination reached by the Sun at the summer solstice.

Tropic of Capricorn – the parallel of 23.5 degrees of latitude south of the Earth's equator that determines the extreme possible southern declination reached by the Sun at the winter solstice.

Tropical month – one revolution of the moon around the Earth relative to the vernal equinox.

Tropical year – one revolution of the Earth around the Sun relative to the vernal equinox.

Troposphere – the first level of the Earth's atmosphere where energy is absorbed and re-radiated to and from the ground.

True vacuum – a state which is devoid of all matter and radiation.

Trumpler classification – a classification of open star clusters.

Tully–Fisher relation – an empirical relationship between the mass or intrinsic luminosity of a spiral galaxy and its asymptotic rotation velocity or emission line width.

Tunguska event – a large explosion that occurred near the Stony Tunguska River in Yeniseysk Governorate, Russia, on the morning of 30 June 1908.

Tunneling – the ability of a particle to overcome an energy barrier without having the required energy.

Tuning fork diagram – a graphical diagram of galaxy types proposed by Edwin Hubble.

Twenty–one centimeter line – the emission line of neutral hydrogen in interstellar clouds.

Twilight – the occurrence whereby the brightness of the Sunlight in the sky gradually increases at Sunrise and gradually fades at Sunset.

UBV system – a wide band photometric system for classifying stars according to their colours. It is the first known standardized photoelectric photometric system. The letters U, B, and V stand for ultraviolet, blue, and visual magnitudes, which are measured for a star then two subtractions are performed in a specific order to classify it in the system.

Ultramicroscopic – microscopy of length scales comparable to Planck length.

Ultraviolet catastrophe – an infinite amount of energy spread among the high frequencies of energy radiation (predicted by classical physics but does not exist in nature).

Ultraviolet radiation – electromagnetic waves with a wavelength between visible light and x–rays.

Umbra:

(1) the centre of a shadow created by a celestial body reflecting the light from a powerful source such as the Sun.
(2) the central dark area of a Sunspot.

Uncertainty principle – the concept that their is a limit to how well the position and velocity of an object can be measured.

Understanding the Universe – elaborating the information already collected and trying to erect a frame of reference, eventually arriving at a structure of a theory leading to further developments in this matter.

Unified theory – a tentative amalgamation of the four forces of nature and having them act as one aggregate.

Uniform motion – movement at a constant velocity in a straight line.

Uniform vibrations – the motion of a string which moves without changing its shape.

Uniformitarianism – the concept that changes in the crust of the Earth are due to the natural forces (wind, water) over a long period of time.

Universal time – (GMT - Greenwich Mean Time) the standard Time accepted throughout the world.

Universality – the idea that different groups act in the same way under all circumstances.

Universe – the sum contents of all of space and time.

Unruh effect – the prediction that an accelerating observer will observe blackbody radiation where an inertial observer would observe none.

Unstable – particles that are prone to decay.

Up quark – a first generation quark with two thirds of the elementary charge and together with down quarks make up protons and neutrons.

Vacuum – empty space.

Vacuum energy – the energy believed to reside in empty space. Also called "the cosmological constant".

Vacuum fluctuations – according to the uncertainty principle, the fluctuations that quantum fields should inexorably undergo.

Vacuum genesis – the concept that the Universe was formed out of no matter.

Van Allen belts – two regions of high energy charged particles in the Earth's magnetosphere.

Variable star – a star with variable brightness.

Variation – an irregularity in the moon's orbit, due to the gravitational pull of the Earth.

Velocity – the speed of an object with regards to the direction of motion.

Vernal equinox – the moment the Sun crosses the celestial equator from south to north.

Very large array – an array of radio telescope located in New Mexico, USA.

Very long baseline interferometer – a technique in radio astronomy by which a multitude of signals from telescopes situated far apart, are turned into one sole image e.g. a series of ten 25m radio telescopes spread along the continental USA and Hawaii.

Vibration number – a number which defines the total value of the energy proceeding from a motion.

Vibrational pattern – a total description of the oscillations of a string relative to all its aspects.

Vignetting – a defect of the lens of an optical apparatus which spreads the incoming light unevenly.

Virgo cluster – a cluster of galaxies near the Milky Way.

Virgo super cluster – a huge cluster of approximately 47,000 galaxies including the Milky Way.

Virtual particles – particles appearing briefly, carrying energy and having an effect on the behaviour of other particles in a gravitational field.

Viscosity – a property of a liquid to have different degrees of fluidity.

Visibility horizon – the length of visible distance.

Visible light – radiation that can be seen by the human eye.

Visible spectrum – the range of electromagnetic waves visible to the human eye.

Visible Universe – the extent of the Universe that we can see.

Visual binary – a double star in which both components can be distinguished within an image.

Visual magnitude – a measure of the brightness of a celestial object as seen by an observer on Earth in the visible photometric band.

VLA – Very Large Array.

VLBA – Very Long Baseline Array.

VLBI – Very Long Baseline Interferometer.

VLT – Very Large Telescope: an observatory consisting of four optical telescope located in the Atacama Desert in Chile.

Void – a region of space devoid of galaxies.

Volatile – a substance that evaporates easily under 'normal' physical conditions (e.g. water).

Volcanism – the eruption of magma (molten rock) onto the crust of a planetary body.

W boson – elementary particles involved in the weak interaction.

Wave – a disturbance that transfers energy through matter or space, with little or no associated mass transport.

Wave mechanics – a method of analysis of the behaviour of atomic phenomena with particles represented by wave equations.

Wave packet – a mixture of numerous waves of all sorts encroaching over each other, excluding a tiny region of space where they can represent a particle.

Wave particle duality – a property of quantum mechanics that bodies possess particle–like and wave–like characteristics.

Wave function – a function that satisfies a wave equation and describes the properties of a wave.

Wavelength – the length between two peaks of an oscillation.

Weak force – a feeble force involved in nuclear reactions, much weaker than the electromagnetic force.

Weak gauge boson – a restricted form of the weak force, mostly w and z bosons.

Weak gauge symmetry – gauge symmetry supporting the weak force.

Weak lensing – gravitational lensing that is not strong enough to break the incoming images, just deform them.

Weakly coupled – a weak coupling where the value of the string coupling is less than 1.

Weight – the gravitational force applied on a body.

White dwarf – a stellar remnant with no more nuclear fuel to burnand supported by electron degeneracy pressure.

Whirlpool Galaxy – a gravitationally disturbed galaxy with the appearance of a whirlpool.

Wide–field camera 2 (wfc2) – the second generation camera installed aboard the HST, on the space shuttle servicing mission in 1993.

Wide–field camera 3 (wfc3) – a fourth generation camera installed aboard the HST, during a space shuttle servicing mission in 2009.

Wien's displacement law – a law stating that while the temperature of a black body increases, its peak emitting wavelength becomes shorter.

Wien's distribution law – a formula that depicts the distribution over wavelengths of blackbody radiation.

Wilson effect – the foreshortening of a Sunspot when it is near the Sun's limb, accompanied by the widening of the penumbra on the nearest side of the limb, and the narrowing on the side farthest on the limb.

Wimp – weakly interactive massive particle – a hypothetical atomic particle which was devised in order to account for the missing mass of the Universe.

Winding energy – the energy built in a string that is coiled around a circular dimension of space.

Winding mode – a string mode that lies around a part of space

Windmansttaten pattern – figures of long nickel-iron crystals, found in the octahedrite iron meteorites and some pallasites. They consist of a fine interleaving of kamacite and taenite bands or ribbons called lamellae.

WMAP – Wilkinson Microwave Anisotropy Probe

Wolf–Rayet star – a normal stage in the evolution of very massive stars, in which strong, broad emission lines of helium and nitrogen ("WN" sequence), carbon ("WC" sequence), and oxygen ("WO" sequence) are visible. Due to their strong emission lines they can be identified in nearby galaxies.

World line – a curve in space–time joining the positions of a particle throughout its existence.

World parameter – an illustration of the density change of dark matter in the process of expansion.

World view – the totality of objects composing our world.

World sheet – two dimensional surfaces traced out by strings in their movement through space–time.

Wormhole – an imaginary tunnel in space joining black holes.

W–particles – massive bosons, remnants from the early Universe.

Wrinkle ridge – long, low ridges found on the surface of the Moon.

Z

Z boson – a particle that transmits the weak nuclear force.

Zeeman effect – the splitting of a spectral line into its original components by a strong magnetic field.

Zelenchuskaya observatory – an observatory in the Caucasus Mountains (Russia) on Mont Pastukhov.

Zenital Hourly Rate (ZHR) – the number of meteors a single observer would see in an hour of peak activity, assumed the conditions are excellent (stars visible up to magnitude 6.5).

Zenith – a fictitious point at 90° vertical from the observer.

Zero AgeMain Sequence – the basic sequence in the Hertzsprung–Russell diagram describing stars which have just started evolving.

Zero point motion – the residues of an old quantum system that stays on the firmament due to the uncertainty principle.

Zodiac – A region of the celestial sphere which stretches about 8-9 degrees either side of the ecliptic, and is the region of the sky where we can find the Sun, Moon and planets (except for Pluto). The zodiac is narrow because most of the planets have orbits that are only slightly inclined to that of the Earth. The exception is Pluto, whose orbital inclination of 17 degrees takes it out of the zodiac during part of its orbit.

Zodiacal light – a cone of weak light seen at all times at the tropics when there is no moonlight and arising from the reflection of Sun light off of ice and dust particles in the Solar System.

Zone of avoidance – a zone beyond the Milky Way where external galaxies are missing, absorbed by the interstellar gas and dust in our galaxy.

Zoo hypothesis – the inference that superior intelligences have already revealed themselves to life on Earth, but avoid involvement in our development.

Bibliographic references

Calle I, Carlos, *The Universe*, New York: Prometus Books, 2009

Clegg, Brian, *Before the Big Bang*, New York: St Martin's Press, 2009.

Evelyn, Gates, *Einstein's Telescope*, New York: W. W, Norton & Co, 2010.

Garfinkle, David and Richard Garfinkle, *Three Steps To The Universe*, Chicago: University of Chicago Press, 2003.

Gates, Bill, *The Road Ahead*, New York: Penguin Books, 1995.

Greene, Brian, *The Elegant Universe*, New York: Vintage Books.

Gibbin, John, *The Universe*, New York: Penguin Books, 2008.

Hawking, Stephen, *A Brief History of Time*, New York: Bantam Books, 2010.

Impey, Chris, *How It Ends*, New York: W. W. Norton & Co, 2010,

Lebans, Jim, *Guide To Space*, Toronto: McCelland and Stewart Ltd, 2008.

Maran, P. Stephen and Marschall A. Laurence, *Galileo's Universe*, Dallas: TX: Benbella Books Inc, 2008.

McEvoy, J. P., *A Brief History Of The Universe*, London: Constable & Robinson 2010.

Musser, George, *String Theory*, New York: Penguin Group, 2008.

Potter, Christopher, *You Are Here*, Toronto: Vintage Canada, 2018.

Repcheck, Jack, *Copernicus Secret*, New York, Simon and Schuster Paperbacks 2017.

Steinhard, J Paul, *Endless Universe*, New York: The Double Day Publishing Group, 2007.

Susskind, Leonard, *The Black Hole War*, New York: Black Bay Books, 2009.

Woodruff, John, *Astronomy Dictionary*, Toronto: Firefly Books, Ltd, 2009.

www.ingramcontent.com/pod-product-compliance
Lightning Source LLC
Chambersburg PA
CBHW060031040426
42333CB00042B/2309